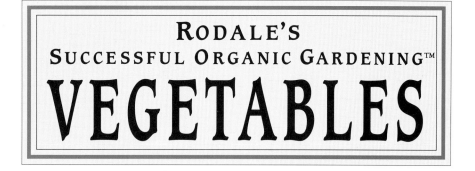

RODALE'S
SUCCESSFUL ORGANIC GARDENING™
VEGETABLES

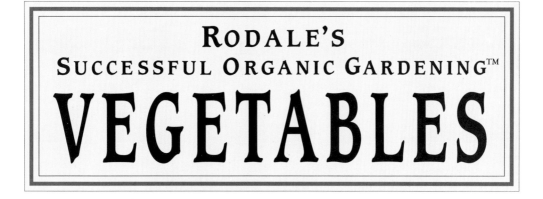

RODALE'S
SUCCESSFUL ORGANIC GARDENING™
VEGETABLES

TEXT BY PATRICIA S. MICHALAK

PLANT BY PLANT GUIDE BY CASS PETERSON

Rodale Press, Emmaus, Pennsylvania

Our Mission

We publish books that empower people's lives.

RODALE BOOKS

If you have any questions or comments concerning this book, please write:

Rodale Press
Book Readers' Service
33 East Minor Street
Emmaus, PA 18098

Library of Congress Cataloging-in-Publication Data

Michalak, Patricia S.
    Vegetables / Patricia S. Michalak and Cass Peterson.
        p.   cm. — (Rodale's successful organic gardening)
    Includes index.
    ISBN 0–87596–563–6 hardcover — ISBN 0–87596–564–4 paperback
    1. Vegetable gardening. 2. Organic gardening. 3. Vegetables.
    I. Peterson, Cass. II. Title. III. Series.
    SB324.3.M53   1993
    635'.0484—dc20   92–31890
                                                  CIP

Printed in the United States of America on acid-free ∞, recycled paper ♻

Rodale Press Staff:
    Executive Editor: Margaret Lydic Balitas
    Senior Editor: Barbara W. Ellis
    Editors: Nancy J. Ondra and Fern M. Bradley
    Copy Editor: Barbara M. Webb

Produced for Rodale Press by Weldon Russell Pty Ltd
107 Union Street, North Sydney NSW 2060, Australia
a member of the Weldon International Group of companies

    Publisher: Elaine Russell
    Publishing Manager: Susan Hurley
    Senior Editor: Ariana Klepac
    Editor: Margaret Whiskin
    Horticultural Consultants: Cheryl Maddocks, Tony Rodd
    Species Charts: Patricia L. Kite
    Copy Editors: Bruce Semler, Yani Silvana,
        Dawn Titmus, Jill Wayment
    Designer: Rowena Sheppard
    Picture Researcher: Anne Nicol
    Photographer: David Wallace
    Illustrators: Barbara Rodanska, Kathie Smith
    Macintosh Layout Artists: Honor Morton, Edwina Ryan
    Indexer: Michael Wyatt
    Production Manager: Dianne Leddy

A KEVIN WELDON PRODUCTION

Distributed in the book trade by St. Martin's Press

    4 6 8 10 9 7 5   hardcover
 2 4 6 8 10 9 7 5 3   paperback

CONTENTS

INTRODUCTION

Most gardeners who have turned to the organic method have done so primarily because of their vegetable gardens. They have found the tastiest vegetables are those that are home-grown organically—rich in vitamins and minerals and free of poisonous sprays.

Gardening organically means becoming aware of the environment and respecting the balance of nature. Chemical sprays and fertilizers can alter this balance by killing off beneficial insects and birds and eventually rendering the soil lifeless. The organic gardener observes the laws of nature and restores plant residues to the soil through mulching and composting. This, in turn, feeds the bacteria, fungi, insects, and earthworms that prepare soil for living plants.

The chapters of this book guide you through all aspects of organic vegetable gardening, beginning with how soil, climate, topography, and exposure determine which vegetables you can grow in your area.

You'll also find information to help you choose the appropriate size and style for your garden. Growing vegetables in rows is the simplest method but there are plenty of alternatives: borders; raised beds; containers; open spaces in paved areas, along paths, and beside walkways; and interplanting with flowers.

Deciding how much to plant and choosing the best cultivars are two of the most vexing issues faced by new vegetable gardeners. You'll learn how to keep garden records, which are invaluable for future planning.

In this book, you'll discover how to shop for the best plants and how to propagate them. The majority of vegetables are easy to grow from seed. You'll learn how you can often get a head start by sowing the seed indoors. You'll also find the facts you need to prepare healthy garden soil to get your young plants off to a good start. Use the information on adding organic nutrients to provide the materials your plants need for vigorous growth. And learn how to recycle garden and kitchen wastes into a valuable soil amendment by creating a compost pile. You will also become aware of how you can prevent and control the occasional pest and disease problem.

To help you get the highest yields out of a limited gardening space, you'll find information on special techniques such as interplanting and succession planting. If you are interested in building raised beds to improve soil drainage, there are complete details on how to create and maintain raised beds. Tips on companion planting will give you the information you need to experiment with this technique in your vegetable garden. And you'll learn lots of tricks to extend your gardening season: You can get an earlier start in the spring and prolong your harvests in the fall by protecting your crops with structures like cold frames, hot beds, and row covers.

For harvesting, you'll learn to identify when vegetables are at their peak of tenderness and flavor and how to handle and store the crop.

The extensive "Plant by Plant Guide," starting on page 78, will ensure success with the vegetables you choose to grow. Each plant entry gives all the details you need to grow productive, healthy plants.

Opposite: Vegetable gardens don't always have to be green and boring to look at. There are many colorful vegetables available, such as ornamental cabbages, that are attractive as well as being edible.

HOW TO USE THIS BOOK

There is nothing more satisfying than growing vegetables. And whether you choose to have a small kitchen garden plot, or you are into vegetable growing in a big way and want to become self-sufficient, the advice in this book will provide information to help get you started and keep your vegetables growing successfully.

Rodale's Successful Organic Gardening: Vegetables is divided into two sections. The first section explains how to select, plant, maintain, and harvest vegetables organically. Read this section carefully if you're a beginning gardener. The second section is the "Plant by Plant Guide," which starts on page 78. It is a guide to common and not-so-common garden vegetables that sets out in convenient reference form the requirements and characteristics of most vegetable species. Consult this section to learn about the specific requirements of each vegetable you grow.

"Understanding Your Garden," starting on page 12, explains how climate, sunlight, and soil influence the growth of your vegetables. Use the USDA Plant Hardiness Zone Map on page 154 to determine the average annual minimum temperature of your particular area. This chapter will also explain topography, as well as the different microclimates that affect growth in your garden and the advantages and disadvantages of these microclimates. And when you understand the topography of your garden, you can choose with confidence the vegetables that are suited to these conditions.

Of course, we know that soil is the key to success with a vegetable garden, and "Understanding Your Garden" fully covers soil composition and requirements so you can improve the soil you already have.

Once you have decided to grow vegetables, the site and size of your plot are important considerations. All the necessary information is covered, including how much to plant and how to keep garden records. There is even information about different garden styles.

When you have done the groundwork, the next step is deciding which vegetables you want to grow. The chapter "Choosing Your Plants," starting on page 30, will familiarize you with the life cycles of each type of vegetable—annuals, biennials, and perennials. This chapter also contains an explanation of the naming system for plants to give you a greater understanding of vegetables. You'll find information that explains the significance of the botanical names of plants, as well as guidelines to help you choose which crops and cultivars to grow. There are tips on choosing the best plants so you will have confidence when you arrive at the nursery to buy seeds and seedlings.

In "Cultivating and Planting," starting on page 40, you'll learn how to prepare your garden for planting. Find out how to have your soil tested for nutrients, and how to correct imbalances, as well as easy ways for you to make your soil productive. There are tips on using green manures, improving soil drainage, and composting—important aspects of creating healthy, organic soil. Ways to make your own compost are fully presented, so you can avoid waste in your kitchen and garden, while at the same time having nutrient-rich material to build and enrich your soil.

Once you've prepared your garden, you may move on to planting your seeds. This chapter takes you through all the steps for successful seed starting, from obtaining the correct containers and growing mediums to sowing, light requirements, thinning, and transplanting.

You will learn how to use companion planting and

interplanting techniques for the best gardening success. Most organic gardeners practice crop rotation for pest prevention and to make the best use of soil nutrients. In "Cultivating and Planting" you will learn these simple techniques. And information on succession planting and successive sowing will show you how to have a continuous harvest through the season. You will also learn about season extension so that you can squeeze some extra time into the growing period.

Once your vegetables are up and growing, the advice given in "Maintaining Your Garden," starting on page 62, will ensure that they remain healthy and pest-free organically. There is advice on monitoring plant growth and interpreting the signs that indicate when it's time to water and how much to water. Practical information is given on mulching as well as ways to train vegetables to save space and make cultivating and harvesting easier. There is a rundown of common pests and diseases and tips on how to identify and control them.

The ultimate reward comes when you harvest your crop. In this chapter you will find out when to pick your crop and how to handle it. You'll also learn about the methods of harvesting, storing, and preserving your vegetables so you can enjoy your produce long after the first frost has appeared.

Plant by Plant Guide

This alphabetical listing, by common names, is an immensely practical and simple reference to a wide variety of vegetables. The Latin (or botanical) and family name is also included for each vegetable.

Each entry is accompanied by a color photograph for easy identification. Topics covered include best climate and site, ideal soil conditions, growing guidelines, pest and disease prevention, common problems, days to maturity, harvesting and storing, special tips, and other common names. Each entry usually describes the main species of vegetable, but also includes information on other species, cultivars, and varieties.

The "Plant by Plant Guide" is designed to help make organic gardening as easy and rewarding as possible by providing all the necessary information in a clear and concise way. The diagram below helps explain what to look for on these practical pages.

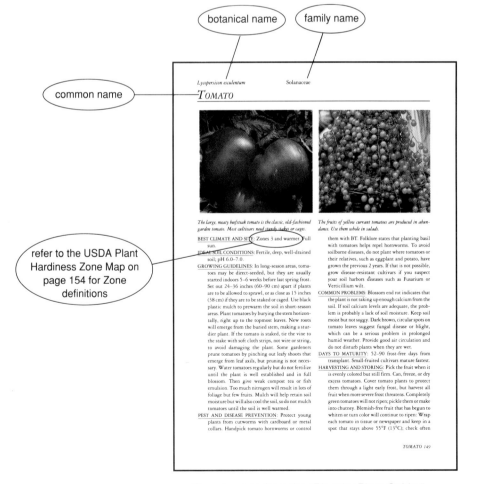

Sample page from the "Plant by Plant Guide."

UNDERSTANDING YOUR GARDEN

Having an understanding of your garden is the first step to success with organic vegetable growing. In this chapter, you'll learn how aspects of the environment, including climate, topography, exposure, and soil, influence the health and productivity of the vegetables you grow.

To a large extent, your climate will determine which vegetables you can grow well. You'll find out how various aspects of climate, such as average high and low temperatures, frost-free period, and rainfall, can influence your choice of crops. You'll also learn how to spot special areas in your garden that may offer slightly different conditions than the open garden, such as warm, sheltered nooks that can give heat-loving crops an extra boost, or lightly shaded spots that can prolong the season for heat-intolerant crops. An awareness of these different microclimates will help you choose the best sites for the crops you want to grow.

Topography, exposure, and soil are other factors to consider when planning your garden. Hilltops, slopes, and valleys all have their unique advantages and disadvantages for gardening. The amount of sunlight a site receives will determine which vegetables you can grow there, so you'll need to keep that in mind as you choose a site for your garden. The quality of the soil is another critical factor to consider for successful garden planning. Healthy soil will yield healthy plants. You'll learn how to identify the physical and chemical characteristics of your soil, and how you can protect or improve the soil with good management practices.

Once you've considered all of these factors, you'll want to put your new knowledge to work, in planning your best vegetable garden ever. After you know where you want to put your garden, you need to decide on how big to make it, depending on the time you have to care for it and the amount of produce you plan to use. Other factors to keep in mind are the style of gardening you plan to use (such as rows or raised beds), the specific crops you want to grow, and how much of each crop you're going to plant.

As you start planning your vegetable garden, you'll begin to appreciate the value of maintaining a good system of garden records. You'll want to keep track of all the things you learn about your particular site, the decisions you make about what and how much to grow, and also the performance of your garden over each season. These sorts of garden records are invaluable reference tools for planning future gardens, and will lead you on to even greater success in growing healthy and flavorful vegetables.

Opposite: When siting your vegetable garden you'll need to take into account a wide range of factors, including climate, topography, soil, and exposure. All of these factors interact to affect the growth of your plants.

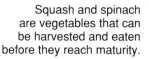

will continue growing, albeit very slowly, even when air temperatures drop as low as 40°F (4°C).

The seeds of cool-season vegetables may germinate in soils as cool as 35 to 40°F (2 to 4°C), but they will sprout most readily in soils that are slightly warmer (40 to 60°F/4 to 15°C). These plants grow quickly and compactly, producing a crop in spring or fall. They tend to make greater nutritional demands on garden soil, so you'll have to make sure that your soil contains plenty of nitrogen and phosphorus for these fast-growing producers. For more information on soil nutrients, see "Soil" on page 22 and "Soil Fertility and Productivity" on page 42. Healthy soil grows healthy plants.

Climate

More than any other single factor, climate influences the selection of vegetables you can grow successfully. Meteorologists use the term *climate* to describe how air temperature, moisture, wind, and other factors interact in specific regions to produce local weather.

The climate of your location also depends on its latitude, altitude, and proximity to mountains and bodies of water. Much of North America falls within the temperate zone, between the extremes of the tropics and the polar circles. The temperate zone is characterized by alternating periods of moderately wet and dry weather, with prevailing winds from the west. Fortunately, most of the vegetables you're familiar with are well suited to this temperate climate.

Within larger temperate zones are "mini-regions," which may differ from the average by being drier, wetter, cooler, or warmer than surrounding areas. If you live in one of these extreme locations, seasonal conditions will be affected and you will have to adapt your plans to meet the rigors of local weather patterns.

No matter where you garden, become familiar with your climate and the factors that influence local weather. Your vegetables have three basic, climate-dependent requirements: a suitable temperature range, a favorable frost-free period, and an adequate supply of moisture. Monitor the weather in your area, and be prepared to take action if your plants aren't getting the conditions they need to grow to their best.

Temperature Range

Most plants have a range of air temperatures in which they grow best. The common garden vegetables are usually divided into two broad groups: "cool season" and "warm season." Cool- and warm-season vegetables vary in their planting and nutritional requirements.

The cool-season vegetables, like asparagus, broccoli, onions, and spinach, are hardy and frost-resistant. They

Okra is a popular warm-weather crop that requires 50 to 60 frost-free days in order to reach maturity.

Cool-weather Crops

Arugula, beet, broccoli, brussels sprouts, cabbage, cauliflower, chicory, Chinese cabbage, collard, corn salad, escarole, kale, kohlrabi, lettuce, mustard, onion, pea, radish, spinach, turnip.

Bean seeds require warm soil to germinate and are an ideal crop for a warm-weather garden. Pole beans take more time to mature than bush beans but bear their crops over a longer period.

Warm-season vegetables, like sweet corn, snap beans, tomatoes, and eggplant, are tender and frost-susceptible. These heat-loving vegetables require an air temperature of at least 50°F (10°C). Serious sweet corn growers intent on harvesting the earliest ears should monitor their soil temperature at planting time, since sweet-corn seeds germinate best at soil temperatures of between 60 and 95°F (15 and 35°C). Seeds of other warm-season vegetables, like snap beans and tomatoes, also prefer warm soil. They are more likely to succumb to soilborne diseases when soil is cool and damp.

Outside their preferred temperature range, vegetable plants may die or become dormant. Dormancy means that the chemical processes that normally occur inside the plant have slowed to a temporary stage of rest. For example, potato tubers left in the ground in fall will remain dormant until the soil warms them the following spring. Some gardeners are able to "overwinter" fall-planted spinach. The young plants remain dormant during the cold season, then resume growing as temperatures rise in the spring. These clever gardeners harvest the earliest crops in their neighborhood.

Warm-weather Crops

Cantaloupe, carrot, corn, cucumber, eggplant, lima bean, okra, parsnip, peanut, pepper, potato, pumpkin, rutabaga, snap bean, summer squash, sweet potato, Swiss chard, tomato, watermelon, winter squash.

Fava beans and cabbage will tolerate some cold and may be planted 4 weeks before the last spring frosts. In cool-summer areas, fast-maturing cultivars of cabbage may be sown successively every month.

Growing Season

In temperate climates, the growing season begins after the last frost of the cold season, and ends when frosts begin again in fall. However, you can begin gardening activity before the growing season technically begins, and continue beyond its end, by using the methods described in "Season Extension" on page 60.

Frost occurs when the air temperature around plants falls below 32°F (0°C) during the night, after dew has formed. The result is a lacy, white coating of water crystals on your plants' leaves the following morning. Although it appears to be limited to the surface, frost damages the interior of the plant as well. Sensitive plants die when exposed to frost in the spring or fall. Hardier vegetables, like most members of the cabbage family, will tolerate some frost damage.

Carrots and radishes are harvested before they reach maturity. They may also be pulled early as baby vegetables.

Corn, eggplants, and tomatoes are sensitive to frost. Wait until the soil is warm and the air temperature is at least 50°F (10°C) before you plant them in the garden.

Obviously, you should wait until the danger of frost has passed before setting out your cold-sensitive plants. If a cold snap threatens after planting, you can protect individual plants by covering them with cloches, or whole beds by draping them with newspaper or lightweight fabric. Read "Season Extension" on page 60 to learn more about protecting vegetables from the cold.

If you are raising a crop for its fruit or seeds, the length of your growing season can be an important factor to consider when planning which vegetables you will grow in your garden. If you garden in an area with a long frost-free season, you can choose from a wide range of crops. But if you live in an area with a short growing season, your choices are more limited. You'll have the best success with fast-growing crops like peas. Crops that need long periods of warm or hot weather to mature, like watermelons, may not produce well in short-season areas. Fortunately, plant breeders are developing faster-maturing cultivars of many heat-loving

crops, so even northern gardeners can enjoy the taste of a vine-ripened watermelon, despite the cooler climate.

A crop that doesn't need to mature (ripen its fruit or seed) before harvest can usually be raised anytime during the frost-free period. You can start harvesting leafy crops like spinach, lettuce, and chard as soon as they are a few inches (5–10 cm) tall. Carrots, radishes, cucumbers, and baby squash are other examples of crops that are always harvested before they are mature. To find out the best harvesting times for your crops, check the seed packets or plant identification labels, or look them up in the "Plant by Plant Guide," starting on page 78.

Moisture

As a general rule, vegetables require at least 1 inch (25 mm) of water each week. In arid regions, or where soils are sandy and drain quickly, 2 or 3 inches (50 or 75 mm) of water may be necessary. Rain is the best source of moisture during the growing season, but you can use irrigation if rains are inadequate.

In some climates, snow contributes significantly to annual precipitation. Even though it falls at a time when most gardens aren't active, melted snow contributes to the supply of water stored below the soil surface.

Drought occurs when moisture levels are consistently inadequate to meet the needs of your plants. High winds and a lack of rainfall can each contribute to the problem of low soil moisture. Without adequate

Chicory 'Rouge di Verone' may be grown as a delicious winter salad crop in areas with a mild climate.

Flowering cabbage provides good fall color in the vegetable or flower garden.

Winds have a drying effect on the vegetable garden. A windbreak made from nylon netting will give protection and not cast too much shade.

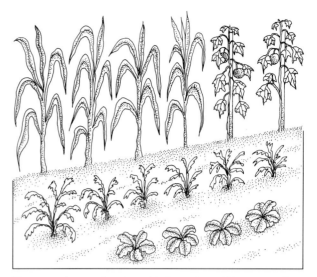

Crops such as spinach and lettuce do not like excessive heat and appreciate the shade cast from tall-growing vegetables like sweet corn.

moisture, most plants will grow slowly, become dormant, or die before they reach maturity. Cool-season crops, and vegetables with shallow root systems (like sweet corn), are most susceptible to low moisture levels. Young seedlings need a more readily available supply of water than mature plants, which usually have extensive root systems. Leafy vegetables that are both drought- and heat-sensitive, like lettuce and spinach, will bolt or wilt when subjected to hot, dry conditions. Bolting, or the premature development of seeds, is the plant's way of ensuring another generation of plants during times of stress. Because most common vegetable crops are fairly susceptible to problems with lack of moisture, it's important to make sure your plants have sufficient water. To learn more about supplying your vegetables with water, read "Watering" on page 66.

Asparagus, broccoli, and onions are all hardy, frost-resistant vegetables.

At the other extreme, excessive moisture reduces the amount of oxygen and nutrients a plant is able to absorb through its root system. Plants susceptible to flooding damage grow slowly, become dormant, or die. A few vegetables, such as celery and watercress, are able to tolerate wet feet. But when choosing a site for your vegetable garden, keep in mind that most other crops prefer moist but well-drained soil.

Watermelons require at least 100 days of warm weather to reach maturity. They are more drought-tolerant than other melons.

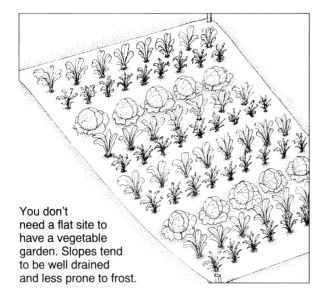

You don't need a flat site to have a vegetable garden. Slopes tend to be well drained and less prone to frost.

Topography

Topography refers to the lay of the land. While a level plot is the ideal gardening site, you will most likely be forced to make do with whatever nature and time have sculpted. Valleys, hilltops, and slopes can provide challenges for the most experienced vegetable gardeners.

Hills and valleys create slight differences in climate, or "microclimates," within the average climate of your region. If you garden on a hill or in a valley, you'll be faced with slightly different conditions from those on the surrounding flat land.

Gardening in a Valley

The air temperature in a valley tends to be cooler than on the flat land above. You can experience this on summer evening bicycle rides when you travel from high to low ground. As you dip into the valley, the air cools. As you leave the valley, the air warms.

This happens because cool air is heavier than warm air. Cool air tends to drain downward and settle on lower ground. For this reason, gardens located at the bottom of valleys suffer the latest frosts in the spring, and the earliest frosts in the fall. The absence of wind in valleys exaggerates this susceptibility to frost.

Soil tends to be wetter in a valley, since water flows down from the surrounding slopes. The water forms puddles, leaving spongy wet areas of ground where drainage is poor. Fungal plant diseases are a greater threat in valleys, since air circulation is poor and moisture levels are high.

You may, however, find the best topsoil at the bottom of a valley. As the water flows downhill, it erodes the slope above and carries loosened topsoil and organic matter down to the valley.

Gardening on a Hilltop

You'll have greater protection from frost and plant disease on a hill, compared with the valley, since air circulation is greater. Excessive winds, however, can cause plant damage, increase soil erosion, and speed up the loss of moisture from the soil.

Since water drains from high to low ground, soil at the top of a hill is drier than the flat land below. On hilltops, soil is thinner due to erosion, and loses more nutrients due to excessive water drainage.

Gardens near water are less prone to late spring and early fall frosts, giving you an extended growing season.

Planting on a slope may boost your harvest. This large vegetable garden, situated on a southern slope, benefits from a longer growing season. It warms early in the spring and remains warm well into fall.

Gardening on a Slope

Locating your vegetable garden on a slope can offer several advantages. While cool air continues moving downward in the evening, the slope remains enveloped in a pocket of warm air that keeps away frost. In cool, northern climates, a slope with a southern exposure warms more quickly in the spring, and remains warm for a longer period of time at the end of the growing season. This means a longer growing season and more harvests.

Slopes, however, are subject to erosion from water drainage and from wind. Excessive water drainage may make it difficult to provide your plants with adequate water. Topsoil may be thin on a slope due to soil erosion. If your site has a fairly steep slope, you may need to build terraces across the slope to hold the soil and keep water from running off too quickly.

The Effect of Water

Bodies of water always influence the climate of the surrounding land. Compared with land, water is slow to heat and slow to cool. The result is that large bodies of water moderate the climate of the land that surrounds them.

If you garden near a lake or the ocean, your microclimate will be different from gardens not located near water. For example, if your garden is near water you many find that frost comes later in the fall than it does to the "average" garden in your USDA Zone. If your garden is located in a valley with a body of water like a stream or a lake, the moderating effect of the water may be enough to hold late and early frosts at bay.

This terraced retaining wall planted out with vegetables demonstrates an economical use of garden space. Fill the terraces with a good compost mix before planting your vegetables, to ensure healthy growth.

Exposure

The intensity and duration of sunlight are critical factors in your vegetable garden. Without adequate sunlight, growth and flowering would cease. Light is your plants' energy source for making food. Most plants prefer sun, but others do better in shade. Light affects plants' size and form, as well as growth rate and yield.

Light Intensity

In climates with frequent cloud cover, the light is said to be less intense than in areas with clear skies. Light intensity influences photosynthesis, which is the plant's internal mechanism for turning light energy into carbohydrates for food, with the help of carbon dioxide and water. Photosynthetic activity increases with greater light intensity, and decreases with less.

Plants vary in the amount of light they require. Some vegetable plants are easily burned by the sun.

When a plant receives too much sun, it develops a bleached appearance and grows poorly. Lettuce leaves will appear papery, tomato fruit becomes whitish and translucent, and cabbage leaves will blister.

With inadequate sunlight, plants become lanky and pale and have poor blossoms and vigor. If plants are illuminated from only one side, they may exhibit "phototropism" and bend toward the light source. This explains why vegetable seedlings, when started indoors on a sunny windowsill, need frequent turning. If not turned regularly, stems, leaves, and flowers will naturally lean toward the light, giving the seedlings a lopsided appearance.

Most vegetable plants require full sun. This means they need an uninterrupted 8- to 12-hour period of unfiltered sunlight. Plants that require partial sun usually need about 5 to 6 hours of direct sunlight, with shade or filtered sun the rest of the day.

Plants that require partial shade must have either filtered, indirect light, or some direct light for less than half of the day and full shade the rest of the time. Plants that require full shade need just that: a solid and dense shade away from light. Few vegetables actually require shade, although cool-season vegetables like lettuce will appreciate protection from the sun when grown during the hottest part of the season. If you prune tomato foliage, leave enough to shade the fruits from the direct sun.

The "Plant by Plant Guide," starting on page 78, will tell you how much light each vegetable plant requires. When planning your garden, take note of where the shadows of trees, buildings, and hills fall throughout the day, to help you decide which vegetables to plant where.

Light Duration

Light duration means the same as day length. At the equator, the daily duration of sunlight is fairly

This vegetable garden is sheltered by a natural windbreak. A combination of tall trees on the outside, and smaller trees and shrubs on the inside, provides a wall of protection and keeps wind damage to a minimum.

Radishes are an easy crop for almost any garden. They will grow well in both dappled and full sunlight.

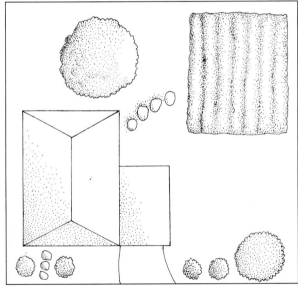

Position the vegetable garden away from tall trees and the house, so their shadows don't affect available sun.

Vegetables Tolerating Dappled Sunlight

Although most vegetables need a full day of sun for good growth, there are a few that can tolerate partial shade. Choose crops that fit your conditions from the list provided below. Arugula, cabbage, corn salad, endive, horseradish, lettuce, pea, radish, rhubarb, spinach, Swiss chard.

Vegetables Requiring Full Sun

If you want to grow a wide range of vegetables, you need to site your garden in the sunniest spot possible. Listed below are some commonly grown vegetables that need full sun. Artichoke, asparagus, bean, beet, broccoli, brussels sprouts, carrot, cauliflower, celeriac, celery, corn, cucumber, eggplant, kale, kohlrabi, leek, melon, mustard, okra, onion, parsnip, peanut, pepper, potato, pumpkin, rutabaga, shallot, summer squash, sweet potato, sunflower, tomato, turnip, watermelon, winter squash.

constant at 12 hours. Moving north or south, light duration tends to decrease but will vary from 8 to 24 hours daily with season and distance from the equator.

The daily pattern of light and dark periods is important as it influences several functions within plants, including seed germination, root initiation, and the growth of blossoms, fruits, tubers, and bulbs. The way that plants respond to light duration is called "photoperiodism." Vegetables such as asparagus, beans, and corn require short days and longer nights in order to bloom and are therefore called "short-day" plants. The "long-day" plants, on the other hand, flower when days are long and nights short. Long-day vegetables include beets, dill, and radish, and most annuals. Other vegetables, like kale, broccoli, cucumbers, and tomatoes, are indifferent to the duration of light they receive and are called "day-neutral."

The day-length needs of some crops, like onions, vary according to cultivar. When choosing cultivars to plant, check to see whether they have any day-length requirements and select the ones that are suitable for your region. You can find this information on the seed packet, in the catalog description, or in the "Plant by Plant Guide."

Pumpkins grow best in full sunlight and moist, well-drained soil.

Good soil structure is the key to a productive garden. Tight, compacted soil (left) is hard to work; plants grow best in loose, crumbly soil (right).

Soil

A good soil is the gardener's key to success. It deserves your first consideration since it influences your plants' growth, health, and yield. Basically, the golden rule with organic gardening is that if you treat your soil well, it will treat your plants well. That means feeding it regularly with organic matter. (See "Composting" on page 46 for more information on feeding your soil.)

Soil Composition

Soil is actually a mixture of minerals, water, air, and organic matter. Average soil contains approximately 45 percent mineral matter, 25 percent air, 25 percent water, and 5 percent organic matter.

Soil Texture Soil texture refers to the fineness or coarseness of your soil's mineral matter. Texture is determined by the relative proportions of sand (very coarse), silt, and clay (very fine); it influences fertility, water retention, and air circulation within the soil. The ideal garden soil is a loam, composed of about 40 percent sand, 40 percent silt, and 20 percent clay. The spaces between soil particles, called pore space, hold water, air, and dissolved nutrients. If the spaces are large, as between sand particles, the soil isn't able to retain sufficient water or nutrients.

Soil Structure The way that the sand, silt, and clay particles join together is called soil structure. Both sand and compacted clays are structureless. In between these extremes, various proportions of sand, silt, clay, and organic matter are arranged as larger pieces, called aggregates. Soil structure is important especially because it influences pore space, which in turn influences water content and drainage, air movement in soil, and the release of nutrients. The ideal soil has a crumbly, granular structure that allows water to drain and oxygen and carbon dioxide to move freely from the air into the

pore spaces. A soil with light, loose structure is said to be friable.

Adding organic matter will improve the structure of all types of soil. See "Adding Organic Nutrients" on page 44 and "Composting" for information on how to improve your soil with organic matter.

Air in the Soil Aeration is critical for healthy soil and vigorous plant growth. Plants take in the oxygen they need from air in the soil. Oxygen is also critical for the well-being of soil organisms, like earthworms. Air also contains gaseous nitrogen, which specialized bacteria convert into a form that your vegetables can use.

Water in the Soil A steady supply of water is critical for the health of almost all vegetable plants. Plant roots absorb water and dissolved minerals from pore spaces in the soil. A good garden soil will hold enough water for plant growth while letting excess water drain down to deeper layers. While most vegetables prefer well-drained soil, there are a few, like celery and watercress, that can tolerate more moisture. Techniques like double-digging and raised-bed gardening can help if you want to improve the drainage of your garden beds; for more details, see "Raised Beds" on page 58.

Organic Matter Soil organic matter refers to the decomposing remains of plants and animals. Organic matter is important because it attracts and holds important plant nutrients in the soil. It is critical for soil structure because it provides the "glue" that holds

---

Acid Tolerance

Vegetable crops vary in their tolerance to acid soil conditions. Match the crops below with the conditions you have available, or amend your soil with lime to raise the pH for the vegetables you want to grow.

Slightly Tolerant (pH 6.0 to 6.8) Asparagus, beet, broccoli, cabbage, cauliflower, celery, Chinese cabbage, cress, leek, lettuce, okra, onion, orach, parsnip, salsify, soybean, spinach, Swiss chard, watercress.

Moderately Tolerant (pH 5.5 to 6.8) Lima bean, brussels sprouts, carrot, collard, corn, cucumber, eggplant, garlic, horseradish, kale, kohlrabi, mustard, pea, pepper, pumpkin, radish, rutabaga, squash, tomato, turnip.

Very Tolerant (pH 5.0 to 6.8) Chicory, dandelion, endive, fennel, potato, shallot, sorrel, sweet potato.

Soil acidity
or alkalinity
affects the
availability
of nutrients for
uptake by plant roots.
Most vegetables
prefer a soil pH of
5.0 to 6.8.

individual particles of minerals together as aggregates. When organic matter has decomposed, its stable remains are called humus. The organic matter content of your soil is influenced by many factors, including climate, vegetation, drainage, soil organisms, and cultivation. Generally, wetter climates and a healthy soil microbial population help create soil with a high organic matter content. Higher annual temperatures, less vegetative growth, and more cultivation lessen the ability of soil to accumulate organic matter. To learn how to add organic matter to soil, see "Adding Organic Nutrients" and "Composting."

Soil Organisms

A diverse population of soil organisms is a critical aspect of soil health. Earthworms are probably the most easily visible indicators of healthy soil. As they burrow through the soil, their tunnels allow air and water easy access to the root zone. And the organic materials that the earthworms excrete, called worm castings, provide a readily available source of nutrients to your plants. Soil microorganisms (plants and animals that are too small to be seen without a microscope) are also important for maintaining healthy soil. They help to decay organic matter and turn it into nutrients that your plants can use. All of these organisms thrive in the same conditions that are preferred by plant roots: a loose, well-drained soil with adequate water and plenty of organic matter.

Soil Nutrients

The amount and type of soil nutrients that are available to your plants depends on the interaction of many factors, including soil texture, structure, moisture, organic matter, and pH. Fine texture, loose structure, ample moisture, high organic matter content, and near neutral pH are all conditions that make the most nutrients available to your plants.

To check the nutrient content of your soil, it may be worthwhile to have a lab test your soil before you plant your crops. It's often a good idea to test the soil in a newly created bed, or in a bed that has been producing for many years. You can get your soil tested by your local Cooperative Extension office or a private soil testing lab. They will provide you with a statement indicating any nutrient deficiencies or excesses, along with recommendations of fertilizers to add to correct the imbalances. Make sure you ask for recommendations for *organic* fertilizers. You may want to test the soil again in a few years to monitor the results of your gardening practices and make any necessary adjustments. For more details on when and how to take soil samples, see "Testing Your Soil" on page 42.

Soil pH

Another useful thing to know about your soil is its pH. Soil pH is a measurement of the acidity or alkalinity of the soil, and is indicated by a range of numbers from 1.0 to 14.0, where 7.0 indicates neutrality. A soil pH of less than 7.0 indicates an acid soil; higher than 7.0, the soil is alkaline. Soil pH is important since it influences soil chemistry. Many more nutrients are available to plants when the soil pH is in the neutral range than when it is very acid or very alkaline. Nitrogen and sulfur become unavailable to plants as the pH drops far below 7.0. Iron and magnesium are unavailable as the pH moves above 7.0.

Many plants prefer a certain pH. Most of the vegetables you grow will do best with a soil pH of 5.0 to 6.8, although many of these will tolerate a pH as high as 7.6. Refer to the "Plant by Plant Guide," starting on page 78, to see what conditions your crops prefer.

If you have your garden soil tested, the laboratory will indicate your soil pH among the results. It should also tell you how much lime to add if your soil pH is too acid for the vegetables you plan to grow. If your soil pH is too alkaline, you can lower it by adding peat moss or sulfur. To find out how and where to have your soil tested, see "Nutrient Analysis" on page 42.

Whether you're raising or lowering your soil pH, the rate of material to add depends on the particular needs of your soil. Follow the recommendations of your soil-testing laboratory, or consult with your local Cooperative Extension Service.

Garden Planning

There are several factors to consider when you're planning a new vegetable garden, or reorganizing an existing one. These factors include your garden's location, size, shape, and design. Consider also your requirements for fresh produce. You may require only a small vegetable garden near the kitchen that will meet your day-to-day demands for organically grown vegetables. Or you may decide to dedicate a substantial area of your yard to vegetable growing so that you'll have enough produce to freeze or store for later consumption.

Site Selection

The first decision to make is where to locate your vegetable garden. You may want to reassign a section of your existing flower garden, or perhaps create a whole new garden devoted entirely to vegetables. Consider the site carefully, because selecting a good location now can avoid many problems later.

Let your garden site influence your selection of plants. Maybe you will have to choose vegetables that prefer shade, or species that can cope with high moisture levels for most of the year. If you have a range of possible sites, choose the one that most closely matches the requirements of your favorite crops. Refer to the "Plant by Plant Guide," starting on page 78, for details about the needs of the vegetables you choose.

Light Most vegetable plants need 6 to 12 hours of sunlight each day. Pick a site well away from shade trees and buildings that cast the garden in shade for more than half the day. If you have a city garden nestled between tall buildings, you may be limited to shade-tolerant vegetables. Gardens located to the south or west of houses or slopes will receive the most light. With such a favorable position, you can start planting as much as 2 weeks earlier, since the soil warms faster.

Moisture Since most vegetables need at least 1 inch (25 mm) of water each week, be prepared to water during dry periods. Make sure you can reach the site with a garden hose or irrigation pipe, or be prepared to carry water to the site by hand. Avoid gardening in low areas that are apt to be flooded, as well as on high areas with excessive drainage.

Soil You'll have the best success if you start with a suitable soil. Avoid heavy clays and loose sands, or amend them with plenty of organic materials (like compost) before you start. Have the soil tested before you begin; adjust pH and nutrient levels the season before, if possible. For more information on how to prepare your soil, see "Soil Fertility and Productivity" on page 42.

Avoid areas that have had heavy foot or vehicle traffic, because this activity compacts the soil and damages its structure. Testing your soil for pollutants is an unnecessary expense unless you suspect the presence of certain contaminants. Urban soils may be polluted by dump sites, industrial waste, motor

It is worth the effort to work out an initial plan for your vegetable garden. Factors to take into consideration include your garden's location, size, shape, and design.

vehicle exhaust, leaded paint, and chemical insecticides. If you do plan to grow vegetables in a city garden, contact your soil-testing laboratory to see if it tests for toxic materials, and learn how to collect the samples required for investigative testing.

Topography Slopes and contours influence the way that you arrange your vegetable plants. Your goal is to minimize the loss of soil through erosion and maximize water retention. If you have to plant on a slope, place your rows or beds across the slope. If your rows of vegetables are arranged running up and down the slope, water will run down the slope instead of penetrating the soil to reach plant roots, and will also carry the precious topsoil away with it.

You'll also want to avoid gardening in low-lying areas, where poor air circulation makes your plants more susceptible to disease.

Obstacles and Access Locate your garden in a place that's easily accessible, both to you and to vehicles that will deliver materials you may need later, like soil or compost. Paths should be wide enough for a wheelbarrow or cart.

As you choose a site, investigate for hidden obstacles like tree roots, shallow boulders, septic systems, and utility lines, and avoid these problem areas.

Time When planning your garden, think about how much time you will have available to spend in it. The amount of work your garden requires will depend on size and will vary with climate and seasons. Don't build a garden that is too big for you to look after. The work will be faster and easier if you have modern equipment, such as a reliable rotary tiller, than if you're using "elbow grease," a shovel, and heavy garden gloves. To save time and energy, you may want to spread the heavy work, like digging new beds or making paths, over several years.

Finances If necessary, you can spread the cost of making your garden over several years. Although

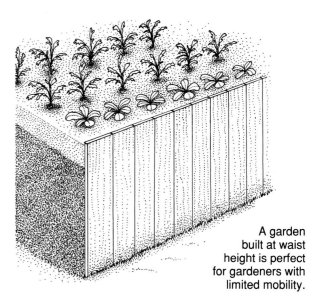

A garden built at waist height is perfect for gardeners with limited mobility.

equipment such as rotary tillers can lighten the workload, it's not necessary to buy expensive equipment at the outset. Start small and enlarge your garden and your equipment collection over time. Your garden plan may include elaborate designs, but you don't need to accomplish everything in one season. If your soil is poor and dry, concentrate on the soil first, and reclaim and improve a new section each season. When vegetable production is steady and you're satisfied with the site, you can concentrate on other aspects, like planting flowers to make your vegetable garden look more attractive, or enlarging your workspace with structures like cold frames or a toolshed.

Special Requirements

Look for a site that fits your individual needs. If your mobility is limited, make sure your garden is easy to reach—for example, just outside the back door. If you're in a wheelchair, you can have a smooth, cement path built and then sculpt a waist-high garden from a slope, or build up a terrace that you can reach easily from a chair. Chair-side gardening is easy if you grow vegetables in tall, recycled garbage cans spruced up with some paint and filled with potting soil.

Vegetables planted in tall, recycled garbage cans and similar containers are easily accessible to chair-side gardeners.

Vegetables for Garden Color

Add excitement to your vegetable garden by growing crops that are attractive as well as good-tasting. Listed below are just a few suggestions of colorful crops and cultivars.

Beans (Stringless) 'Royal Burgundy', 'Royal Purple'.

Broccoli 'Purple Sprouting'.

Cabbage 'Mammoth Red', 'Preko'.

Kale 'Peacock Pink', 'Cherry Gateau'.

Ornamental Cabbage 'Dynasty Pink'.

Peppers *Light green*: 'Lady Belle', 'Yolo Wonder', 'Big Bertha', 'Park's Whopper', 'Pro Bell II', 'Ace'; *Golden*: 'Golden Bell', 'Golden California Wonder', 'Golden Summer'; *Yellow*: 'Gypsy', 'Banana', 'Hungarian Wax'; *Orange and red*: 'Tasty Hybrid', 'California Wonder', 'Tequila Sunrise', 'Banana Supreme', 'Anaheim'.

Squash 'Burpee Golden', 'Gold Rush', 'Jersey Golden', 'Baby Blue'.

Tomatoes 'Golden Boy', 'Yellow Plum', 'Yellow Cherry', 'Yellow Pear'.

Interplant your vegetables with other flowering plants and herbs for added color, and also to save space.

Garden Style

The style of your garden will depend on how much space and time you have. You can garden in rows or beds, both of which have different advantages.

Rows

Row gardening involves planting vegetables in parallel lines. Rows might be the best way to garden if you have plenty of space and time for cultivating between rows. You'll find rows easy to plan and easy to plant, especially for large crops of tomatoes, beans, or corn. You'll use space less efficiently, however, since you'll need to have paths between rows. That means more water consumption and less yield per unit of area, when compared with gardening in beds.

If you plant double instead of single rows, you can squeeze more vegetable plants into your plot. Stagger the spacing when you plant seeds or seedlings, and you'll save time and improve yield. Just think of the two rows as one continuous "zigzag." See "Planting Outdoors" on page 50 for instructions on how to make rows.

Beds

Beds are raised areas for planting, and offer several advantages over row planting. Production is greater, since beds are more intensively planted and aren't subjected to compaction from foot traffic. They're attractive, space-efficient, and easy to organize and work within, and they require less weeding than rows.

Most garden beds are 3 to 5 feet (0.9 to 1.5 m) wide,

Make sure you have paths between beds for easy access. Herbs are ideal to plant along the edges.

Container gardening is an effective way of utilizing space in your garden. It also has a decorative effect.

If you have space, you may grow your vegetable beds in an interesting pattern for a dramatic look in your garden.

with paths of 2 to 3 feet (60 to 90 cm) between them. Make them as long as necessary. For instructions on making garden beds, see "Raised Beds" on page 58.

Garden Size

The size of your garden will be determined by how much time and money you have to spend on it, and how much food you want to produce. New gardeners would be wise to start small, perhaps with one 3-foot by 5- or 10-foot (90 cm by 1.5 or 3 m) bed. The final size of your vegetable garden also depends on the crops you want to grow. Refer to "Growing Habit" in the "Plant by Plant Guide," starting on page 78, for the space requirements of the vegetable crops that you wish to grow.

If space is limited, you might want to make plans for several small gardens. Pick sunny pockets of space along a fence, walkway, house, or garage. You can manage these garden islands as you would a raised bed. For added color, include flowering annuals like marigolds and nasturtiums that won't compete aggressively with your vegetables. Techniques like interplanting and succession planting can also help you get high vegetable yields from a small space; for more details on these techniques, see "Interplanting" on page 54 and "Succession Planting" on page 55.

Container Gardening

You can grow some common garden vegetables, like potatoes, in containers. You can also grow dwarf cultivars of larger-sized vegetables in pots and barrels. Some newly introduced cultivars of tomato, cucumber, pepper, and squash are ideal for growing in containers—they save space *and* produce heavily.

Be sure to choose containers with holes in the bottom for adequate drainage. The size of container you select depends on what vegetable you plan to grow. Small pots are fine for hanging baskets of ornamentals, but may not provide the best environment for heavy-producing vegetable plants. You can recycle waste-baskets, garbage cans, and even heavy plastic bags for use as container gardens. Place the containers on homemade platforms with casters, for greater maneuverability.

This compact vegetable garden is the perfect size to feed a small family.

Plant Selection

Before selecting cultivars, you need to think about how much you want your plants to yield, which cultivars will be best suited to your garden, how much time you're prepared to spend dealing with pests, and what qualities are important to you in your vegetables.

Yield If you have a large family to feed, you might prefer a cultivar that has a high yield. You might choose a less productive cultivar if some other factor, like flavor or pest tolerance, has greater priority.

Growing Requirements When you're trying to decide among cultivars offered by seed suppliers, consider how well they're likely to do in your garden. The pH and nutritive requirements are important, as well as light and moisture needs. Check for frost resistance, hardiness, and the number of days to maturity. Pick one or several cultivars that fit your planting schedule, especially if you plan to make successive sowings or use season extenders. (For an explanation of these terms see "Successive Sowing" on page 55 and "Season Extension" on page 60.) Whenever possible, use cultivars that have a good local reputation. If a cultivar of string bean has done well for your neighbor's family over several years, chances are it will do just as well for you.

Pest Resistance Choose insect- and disease-resistant cultivars whenever possible. These are plants that have been bred specifically for their pest resistance. Older cultivars that were selected before the widespread use of synthetic pesticides may also offer some pest resistance. Check the plant descriptions in seed

catalogs or the "Plant by Plant Guide," starting on page 78, to see whether resistant cultivars are available for the crops you want to grow.

Other Considerations Most gardeners want high yields, but there are also other factors that you'll want to keep in mind when selecting which vegetables and cultivars you want to grow. Flavor, for example, is a top priority with many growers. Heirloom cultivars in particular are known for their "true" flavors, although many new cultivars also offer good flavor.

Another thing you might want to consider is the ease of harvest. Harvesting can quickly turn from a pleasure to a chore if your crops are difficult to pick. Vegetables like snap beans are easier to pick if the pods are clustered at the top of the plant, instead of at the bottom. If you have sensitive skin, prickly vines and leaves may cause a rash, so look for "hairless" or "spineless" cultivars of squash, okra, and other vegetables. If you have trouble

Garden Records

Once you have chosen a site, style, and size for your garden, it's a good idea to draw your plan on paper before you go out and start digging.

Use graph paper marked to the appropriate scale (1 square foot [900 sq cm] per square works well). Mark the outline of your garden first, then add permanent features like trees and paths.

Next, make a list of vegetables you want to grow and then arrange them on your graph paper in rows or beds. Keep in mind their planting and harvest dates, size at maturity, and special growth requirements. Use the "Plant by Plant Guide" to help you.

As the season progresses, keep track of planting dates, pest problems, fertilizer rates, and harvests. Some gardeners make notes in a diary or on a calendar. You can create your own tables of garden data using column headings like planting date, harvest date, yield, and quality. As well as monitoring your vegetable garden's day-to-day progress, your garden records will also help you plan for greater success with the next harvest. For example, by keeping a note of planting dates and harvest, you can record how well different cultivars perform in your garden, or when space will become available for replanting. Update your map and other records with each new gardening season.

bending over, you may want to choose vining cultivars that you can train up a fence or wall, putting the harvest within arm's reach.

Keep in mind that the ripening and harvesting period for each vegetable varies. Commercial growers find it easier to harvest cultivars that ripen all at once, but a home gardener may prefer to spread out the harvest with a cultivar that ripens over a period of several weeks. You'll also want to consider how much time you are willing to contribute to harvesting. Some crops, like corn and snap beans, require daily harvesting to get produce at the peak of quality. Others, like lettuce, onions, and carrots, mature more slowly, so you wait several days between harvests.

Post-harvest preparations, like washing, peeling, or slicing, can be simple or complex. Some cultivars are easier to use than others. Cylindrical-shaped beets and carrots, for example, are easier to slice than round ones. Nutritional quality can vary with cultivar. Some new cultivars claim higher nutritional value than others. Certain carrot cultivars, for example, have more beta carotene than others.

Storage quality is important if you don't use vegetables immediately. Onion cultivars differ in holding quality when stored during winter. And most vegetables vary in freezing, canning, and dehydrating quality. To find out which vegetables are best suited for storage, check out "Harvesting and Storing" on page 76.

How Much to Plant

Experience is the best tool when it's time to decide how much to plant. After only one season of gardening, you can decide to grow more or less than last year. Yield, however, can go up or down seasonally, depending, for example, on local weather and pest outbreaks. That's why many gardeners plant more than they need if they have the space to do it.

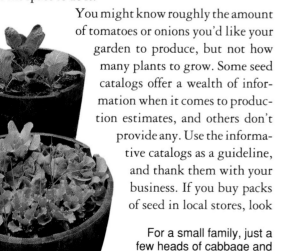

You might know roughly the amount of tomatoes or onions you'd like your garden to produce, but not how many plants to grow. Some seed catalogs offer a wealth of information when it comes to production estimates, and others don't provide any. Use the informative catalogs as a guideline, and thank them with your business. If you buy packs of seed in local stores, look

For a small family, just a few heads of cabbage and a little patch of lettuce can provide plenty of produce.

Before planting a high yield garden, take into account how much produce you will actually use, to avoid waste.

to the planting information on the back for an idea of yield and when to expect it.

The following list will give you things to keep in mind as to how much to plant.

- How much produce will you use fresh, canned, frozen, and dried? Even single gardeners tend large gardens if they're growing for the whole year. A garden planted for fresh use is a pretty small affair. Use your seed source as a guide for how much yield to expect, and figure accordingly.

- When do you want to harvest the crop? Like the preceding question, this depends on how you plan to use it. For example, 1 pound (0.5 kg) of bean seeds will give you a lot of beans at one time, or smaller quantities if you stagger the planting dates throughout the season. Instead of 1 pound (0.5 kg) of one cultivar, you could select smaller amounts of several cultivars with various maturity times for a longer harvest period.

- How much growing space do you have? You can stretch your growing space with raised beds, trellises, succession planting, containers, and other techniques (which you can learn about in the chapter "Cultivating and Planting," starting on page 40). Check your vegetable seed source for the most specific plant spacing recommendations, or use the information in the "Plant by Plant Guide."

- How much storage space do you have? If you are growing crops like potatoes and onions for winter storage, try to avoid planting more than you'll have room to keep. Check the "Plant by Plant Guide" for tips on successfully storing your harvest.

CHOOSING YOUR PLANTS

If you try to imagine the number and variety of plants grown worldwide, you'll soon recognize why it's necessary to classify plants into specialized groupings. And having an understanding of this is important if you want to grow healthy vegetables. Being well informed helps you to choose plants that will make your vegetable garden a success.

In this chapter you'll learn about the origins of your vegetables. For example, you'll learn that vegetables which originate in a hot climate may require special treatment if you want to grow them in a cold climate. This awareness will help you to choose the most reliable vegetables for your particular climate.

Most of the vegetables that you cultivate in your garden will be annuals, which will grow, flower, and set seed in just one season. But there are also biennial and perennial vegetables, which have different life cycles and different needs. In this chapter, you will learn to appreciate the value of understanding these different kinds of life cycles, as it is essential in helping you to make the best choices.

The first key to success in vegetable gardening is to start with healthy plants. This chapter tells you everything you need to know to choose strong and well-adapted plants and seedlings. You'll also find special tips to help you if you are buying seeds instead of seedlings. With this knowledge at your fingertips you can be confident that you are making the correct choice and that you will be well rewarded.

This chapter also includes species charts that will guide you when you are selecting vegetables for your garden. They list vegetables under particular categories, useful for the vegetable gardener when deciding which vegetables to grow. The charts include "Foolproof" Vegetables, Container Vegetables, Miniature and Baby Vegetables, Salad Vegetables, a vegetable Vitamin Guide, Short-season Vegetables, Perennial Vegetables, and Heirloom Vegetables.

Opposite: It is important to consider the family relationships of your vegetable plants so you don't grow related plants in the same spot year after year. Leeks, for example, are closely related to onions and garlic.

What's in a Name?

Families are groups of plants that share a common ancestor and some characteristics. For example, the Solanaceae family contains potatoes, peppers, eggplants, and tomatoes. A family is composed of genera (singular "genus"), which are more narrowly defined groups of plants and often have a recent common ancestor. The genus name is the first part of a plant's botanical name. The genus *Brassica* contains such vegetables as broccoli, brussels sprouts, cauliflower, cabbage, turnip, and many others. The species name is composed of two parts: the genus name and the specific epithet. *Cynara scolymus*, for example, is the botanical name for the species we know as artichoke.

Sometimes the species name is followed by a third name in italics, a subspecies name, which indicates a group of plants that has developed distinct differences from the main population. For example, the *"sativa"* of *Eruca vesicaria sativa* refers to the subspecies of *Eruca vesicaria* known as arugula.

Plants can be classified into categories like varieties and cultivars. A variety, such as *Apium graveolens* var. *rapaceum* (celeriac), denotes a group of plants that occurs naturally within a species. It has characteristics that distinguish it from the typical species and other varieties within the same species. A cultivar denotes a human-made or human-propagated selection from within a species. Cultivars can be formed either by cross-pollinating two or more species, or by propagating a mutant specimen with unusual characteristics. A cultivar name is listed after the species name and is set off by single quotation marks. Cultivar names often do double duty as the common name. *Zea mays* var. *praecox* 'Strawberry' is often simply called strawberry popcorn. *Zea mays* var. *rugosa* 'Silver Queen' is the well-known, late-season cultivar of sweet corn.

When plant breeders cross two genetically different plants, the result is a hybrid. Hybrids often have superior characteristics, like enhanced resistance to pests or improved flavor. Unfortunately, their offspring are quite variable and do not always resemble the parent plants, meaning that you must purchase new seeds each season. Hybrids of ornamental plants are usually indicated by an "x" in their botanical name (as in *Iris* x *germanica*); vegetable hybrids are usually indicated by their cultivar name (as in 'Burpee's Supersteak Hybrid' tomato).

A hybrid tomato like 'Pink Delight' is a result of a cross between two genetically different plants to create a superior plant.

Peppers, along with potatoes, eggplants, and tomatoes, are members of the Solanaceae family.

Longevity

Most of the vegetables you'll grow are annuals. Annuals germinate, grow, flower, and set seed in one season. Hardy annuals, like spinach and lettuce, tolerate frost and are able to self-sow. (This means that if the plant is left in the garden to make and release seeds, and the right conditions exist for germination, new plants will sprout the following season.)

Half-hardy annuals, also known as "winter annuals," need warm soil for germination but can tolerate light frosts. Half-hardy annuals, like mustard, can be planted in the fall for harvest the following season. Tender annuals need warm soil for germination and are frost-susceptible. Sweet corn is the best-known tender annual. To grow tender annuals in cold climates, you'll need to sow seeds indoors and wait until after the last frost to move them outdoors, or else wait until the soil is warm and the danger of frost has passed to sow seed directly.

Most annual vegetables require full sun and are shallow-rooted. This means they'll require plenty of water, since their roots remain close to the surface.

Biennials, like carrots, are plants that require 2 years to complete their life cycle. During the first year they germinate, then form a small and compact vegetative plant that remains dormant over winter. Biennials depend on the stored nutrients in their taproots to survive the cold season. In the second year, they bloom and make seed, then die at the end of the season. It's best to sow biennials directly outdoors, since their taproot is sensitive to the disturbance of transplanting.

Perennials live for more than two seasons. Short-lived perennials die after 3 to 5 years. Long-lived perennials, like asparagus, live for more than 5 years. A tender perennial is one that survives for more than one season in warm climates but is cold-sensitive in cool climates. Since tender perennials are killed by winter temperatures, they're often grown and referred to as annuals. Some species of pepper are tender perennials in warm climates, even though they're managed like annuals in cool-climate gardens. Most perennials have an extensive and deep root system that makes them less susceptible to drought.

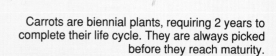

Carrots are biennial plants, requiring 2 years to complete their life cycle. They are always picked before they reach maturity.

Asparagus is a perennial vegetable that will grow in cold climates. It may continue to yield for up to 20 years.

Lettuce, an annual plant, will finish its growth cycle in one season.

Turnips are hardy, annual plants. Fast-growing and easy-storing, turnips thrive in cool temperatures.

Buying Plants and Seeds

Make a list of the vegetables and cultivars you plan to grow, along with an estimate of the number of plants or amount of seed for each. Take your shopping list with you to the garden center, or refer to it as you make out your seed orders from catalogs. By having a list, you'll be more likely to buy just what you need, and you'll avoid ending up with extra plants that you don't have room for.

Always check the labels before you buy. Broccoli and cabbage, for example, look different as adults, but their seedlings are very similar.

Buying Plants

When buying vegetable seedlings, look for healthy, green plants with lush foliage. Reject tall, leggy seedlings in favor of short, full plants. Upright, annual vegetables that are planted individually, like peppers, tomatoes, or broccoli, should be limited to one stem per pot or cell. Extra seedlings may look like a bargain, but it's easy to damage both plants if you attempt to separate them, and the vigor of both plants will be compromised if you plant them together. Pots of vegetables like cucumbers or pumpkins, which are usually planted in hills, can hold only a few small seedlings. Leafy crops like arugula or leaf lettuce aren't as fussy about plant spacing; they can be started, planted, and harvested in clumps.

The soil in the seedling pots should be moist but not soggy. Roots should be well developed, but not so much that they've filled the pot. It's normal for a few threads of root to escape the pot, but avoid transplants with solid masses of tightly woven roots. Also check the seedlings for signs of insect pests or disease symptoms. The presence of beneficial insects, like ladybugs, might

These young squash plants only have their seed leaves. Wait until they have one or two true leaves before you plant them out into the garden.

Buying Seeds

If you're buying seeds instead of plants, you'll need to consider several different factors. First of all, check the date on the seed packet, and only buy fresh seeds. Buy only enough for this year, unless you're able to store seeds under the proper conditions of low moisture and low temperature.

Check the seed packets for germination qualifiers. The United States Department of Agriculture regulates germination specifications, so you're assured a minimum quality. Most reputable seed suppliers exceed the minimum germination standards. A seed supplier who finds that a particular batch of seeds has poor germination might be forced to offer it if substitutes aren't available. If so, the supplier will state on the label that germination is less than optimum.

Before you buy your seeds, it's a good idea to compare prices among several different catalogs or suppliers. As you shop, make sure that the prices you compare are actually for the same amount of seeds. Seeds are sold by weight or number of seeds. Check to see how much you're getting in each packet you buy. Shopping around is time-consuming but worthwhile. You'll also want to look for a reputable dealer. There can be a big difference in seed quality among suppliers—ask any greenhouse manager. Some seeds may germinate poorly, or be infected with seed-borne diseases. The best seeds have excellent germination and plant performance. Talk to other gardeners in your area to find out which suppliers they prefer.

Don't buy crowded seedlings, since the damage caused when separating them will slow their growth.

Seedlings should be well spaced and limited to one stem per pot to allow unhindered and steady growth.

Learn to recognize the difference between a healthy and unhealthy seedling. The healthy tomato plant has rich green foliage, while the tomato with yellow foliage has not been given enough light.

indicate a recent pest problem, but it's better to take home the beneficials than the residue of pesticides. Look for webs, insect droppings, or signs of damage like ragged holes or spotty leaves. Pass up seedlings with discolored or wilted leaves. Look for identification

Learn to identify healthy seedlings for a bountiful harvest.

labels, especially if you're unfamiliar with the seedling stages of vegetables. Closely related vegetables like broccoli, cauliflower, and cabbage, for example, look alike at this stage. If you're interested in a particular cultivar, shop early for the best selection, or you might have to settle for unfamiliar cultivars. Most greenhouse growers have to limit the number of cultivars they sow, since space is limited. Take along a current seed catalog when you shop, and consult it if you have to decide between unknown cultivars. A good greenhouse salesperson should be able to recommend the best cultivar for your garden.

It's a good idea to inquire about how the seedlings were grown. If you're strictly organic, look for growers who use starting medium, seeds, fertilizers, and pest-control practices that conform to organic standards. If you can't find a source for organically grown seedlings and don't want to raise your own, you may have to settle for non-organic seedlings. At this stage, the amount of synthetic materials imported to your garden is most likely negligible.

You'll also want to ask whether the seedlings have been hardened off (gradually acclimatized to outdoor conditions). If the seedlings you buy have been in a warm greenhouse, you'll have to introduce them slowly to the cool outdoors. Read "Hardening Off" on page 49 to learn how to do this.

Once you get your seedlings home, make sure that you keep them watered if you are unable to transplant them to the garden immediately. Place them out of the direct sun. Keep a close eye on the weather, and be ready to bring them indoors if a thunderstorm or frost threatens.

Species Charts

The following species charts are designed to give you an idea of vegetables that are useful or good in particular ways. For example, if you are a beginner at vegetable gardening, there is a list of "foolproof" vegetables that are quite easy to grow. If you have a small garden, it would be useful to know which vegetables make good container plants, in order to be able to make the best use of available space, so there is a chart listing suitable crops. There is also a list of miniature and baby vegetables, which are either naturally small vegetables or vegetables that can be harvested when immature, and are a treat for gourmet cooks. There is also a chart of more unusual vegetables that can jazz up the humble salad. If you are keen on making sure that your diet contains all the right vitamins, there is a vitamin guide to vegetables that lists several vegetables that are high in vitamins A and C. If you live in an area with a short growing season, there is a list of short-season vegetables that are ideal for your situation. There is also a chart of perennial vegetables, crops that you plant just once and then harvest year after year. Finally, there is also a list of heirloom vegetables, which are said to have a better flavor than other types of vegetables, and often have some historical significance or even an interesting story behind their existence.

"Foolproof" Vegetables

To be honest, there really is no such thing as a "foolproof" vegetable—all vegetables need good soil preparation and regular care. But there are some relatively easy-to-grow vegetable crops that are good confidence-builders for the beginning vegetable gardener, or reliable standbys for the more experienced gardener. Some of the most dependable vegetable crops are listed below, along with specific growing hints.

Bean, bush: Pick regularly to prolong harvest

Bean, pole: Needs a trellis

Beet: Keep soil evenly moist

Garlic: Plant in fall to harvest next summer

Lettuce: Grows best in cool weather

Onion: Grow from sets (immature bulbs)

Pepper, sweet: Start plants indoors

Radish: Easy from spring sowings

Squash, summer: Add compost to the soil

Tomato, cherry: Needs ample water

Container Vegetables

Even gardeners who have limited space can enjoy the pleasures of vegetable gardening by simply growing their crops in containers. The size of the container needed depends on which plants you want to grow, but it must have drainage holes. Fill the container with a soilless planting mixture, not with garden soil (which will pack down too tightly with repeated watering). When choosing which vegetables to grow, remember that compact or dwarf cultivars are generally the best choice for containers. Listed below are some vegetables that are well adapted to container culture; for more ideas, see "Miniature and Baby Vegetables" on page 37.

Bean (bush types); carrot (short-rooted cultivars); corn salad; cucumber (bush cultivars); eggplant; kale; lettuce; onion; pea (needs a trellis); pepper, sweet; radish; squash, summer (bush cultivars); Swiss chard; tomato; watercress.

Miniature and Baby Vegetables

For beginning gardeners, the goal is often to grow the biggest vegetables possible—colossal carrots, enormous eggplants, or mammoth melons. Experienced gardeners, though, know that the biggest vegetables don't always have the best flavor. You can enjoy the gourmet treat of miniature or baby vegetables without paying outrageous prices at the grocery store. In some cases, you just grow full-sized cultivars and simply harvest them at an immature stage. Other vegetables have cultivars that naturally produce small fruits. Listed below are hints and suggested cultivars for growing many popular miniature or baby crops.

Bean, snap: Pick pods when they are up to ¼ inch (6 mm) wide; 'La Belle', 'Rapier', 'Vernandon'

Beet: Pull when root is 1 inch (2.5 cm) or more in diameter; 'Baby Gladiator', 'Little Ball'

Carrot: Pull as soon as the roots develop color; 'Caramba', 'Little Finger', 'Planet'

Corn: Pick tiny ears 1 to 2 days after silks appear, or choose small-fruited cultivars; 'Baby Blue', 'Cutie Pops', 'Indian Fingers', 'Wampum'

Cucumber: Pick when fruits are 2 to 6 inches (5 to 15 cm) long; 'Cornichons', 'Lemon', 'Picklebush', 'Saladin'

Eggplant: Pick fruits up to 4 inches (10 cm) long; 'Little Fingers', 'Short Tom'

Lettuce: Harvest small, young leaves as needed; 'Little Gem Mini Romaine', 'Summer Baby Bibb', 'Tom Thumb'

Melon: Choose small-fruited cultivars; 'Bush Star Hybrid', 'Garden Baby', 'Minnesota Midget'

Onion: Pull as needed for scallions, or plant pickling onions for small bulbs; 'Barletta', 'Crystal Wax Pickling', 'Early Aviv'

Pea: Pick when pods are 2 to 3 inches (5 to 7 cm) long; 'Giroy', 'Norli', 'Petit Provencal', 'Precovil'

Pepper, sweet: Pick small peppers as needed; 'Cadice', 'Gypsy', 'Sweet Cherry'

Pumpkin: Choose small-fruited cultivars; 'Baby Boo', 'Baby Pam', 'Jack Be Little', 'Munchkin', 'Spookie'

Squash: Pick summer squash at 4 inches (10 cm) or longer, and choose small-fruited cultivars for winter squash: 'Arlesa', 'Peter Pan', 'Raven', and 'Supersett' for summer squash, 'Sweet Dumpling' for winter squash

Tomato: Choose small-fruited cultivars; 'Camp Joy', 'Chello Yellow', 'Ruby Pearl', 'Sweet 100', 'Yellow Pear'

Salad Vegetables

Salad doesn't just have to be just a bowl of 'Iceberg' lettuce and a slice of tomato—there's a wide variety of vegetables available that can add exciting tastes, textures, and colors to any salad combination. Listed below are some of the more unusual vegetables that can make your salad something extra special.

Arugula: tangy leaves

Bamboo shoot: tender, young shoots

Calendula: colorful flower petals

Celeriac: nutty/celery-flavored roots

Celtuce: crispy leaves

Chicory: tangy leaves

Corn salad: nutty/buttery-flavored leaves

Cress: spicy leaves and stems

Dandelion: slightly bitter leaves

Endive and escarole: tangy leaves

Fennel, Florence: crispy, sweet leaf bases

Jerusalem artichoke: sweet, nutty tubers

Mizuna: lacy, mild leaves

Mustard: spicy leaves

Orach: tender, mild leaves

Purslane: succulent, crispy leaves

Sorrel: tangy, lemony leaves

Spinach: tender, mild leaves

Sprouts: crispy shoots, leaves, and roots

Violet: colorful flowers

Watercress: tangy leaves and stems

Vitamin Guide

Vegetables are a great home-grown source of many vitamins, especially vitamins A and C. The list below shows some vegetables that are particularly high in one or both of these vitamins. Keep in mind that the skins and outer leaves of vegetables are often especially high in vitamins, so use these parts whenever possible. Raw vegetables are highest in vitamins. Heat will destroy many vitamins; cooking water will leach out others, so avoid boiling whenever possible. To preserve the highest possible vitamin content, try methods like steaming and stir-frying that cook vegetables quickly. In many cases, cooking vegetables by microwave gives superior results nutritionally, particularly for those vegetables high in vitamin C.

Vegetable	Vitamin A	Vitamin C
Asparagus	✔	✔
Beet greens	✔	
Broccoli	✔	✔
Brussels sprouts	✔	✔
Cabbage	✔	✔
Carrot	✔	✔
Cauliflower		✔
Endive	✔	
Mustard greens		✔
Pepper, sweet		✔
Pumpkin	✔	✔
Spinach	✔	✔
Squash, winter	✔	✔
Sweet potato	✔	✔
Tomato	✔	✔
Turnip greens	✔	✔
Watercress	✔	✔

Short-season Vegetables

Fast-growing vegetables have many uses in the vegetable garden. If you live in an area with a short growing season, these quick-maturing crops are often your best chance for getting a good harvest. In long-season areas, use these crops before or after a slow-growing crop to get an extra harvest from the same bed. Or interplant short- and long-season crops, such as radishes and cabbage—you'll harvest the fast-growing crop by the time the slower crops need the space. Listed below are several short-season vegetables you can try.

Arugula	Lettuce
Beet	Mizuna
Bean (bush types)	Mustard
Chicory (leafy types)	Orach
Corn salad	Pea
Cress	Radish
Kale	Spinach
Kohlrabi	Swiss chard

Perennial Vegetables

Although most vegetables are annuals—meaning that you have to start new plants each year—there are a few exceptions. Plant perennial vegetables just once, and you'll be able to harvest from them year after year. Listed below are several perennial vegetables, followed by their edible parts. To see if a particular crop will survive over winter in your climate, look it up in the "Plant by Plant Guide," starting on page 78.

Artichoke, globe: flower bud
Asparagus: young shoots
Bamboo: young shoots
Cardoon: peeled stalks
Dandelion: leaves, roots
Daylily: buds, flowers, tuberous roots
Horseradish: roots
Jerusalem artichoke: tubers
Rhubarb: leaf stalks
Sorrel: leaves
Violet: flowers
Watercress: leaves

Heirloom Vegetables

While it's easy to fall prey to the glossy color photos in the spring seed catalogs, keep in mind that the newest vegetable introductions may not always be the best choices for your garden. Many gardeners are finding that heirloom vegetables—plants that have been grown and preserved by backyard gardeners for many decades—often have better flavor than newer, commercially produced cultivars. Others enjoy growing heirloom vegetables for their historical significance; these plants often have fascinating stories behind their existence.

If you are interested in growing heirloom plants, you might want to ask older gardeners in your area whether they have any crops from which they save seeds year after year. Locally grown heirloom crops are often best adapted to the growing conditions in your area. You could also look for catalogs that offer heirloom seeds, or in magazines that feature seed swaps. Listed below are just a few heirloom cultivars for some of the most popular crops.

Bean: 'Borlotto' (shelling bean with rosy red pods), 'Champagne' (vigorous pole bean), 'Fin de Bagnois' (early bush snap bean), 'Green Annelino' and 'Yellow Annelino' (pole beans with crescent-shaped pods), 'Low's Champion' (shelling bean with broad, flat pod), 'Triomphe de Farcy' (bush snap bean with light purple pods)

Beet: 'Chiogga' (early, sweet), 'Long Season' (good for winter storage), 'MacGregor's Favorite' (purple tops)

Broccoli: 'Romanesco' (sweet, conical florets)

Cabbage: 'Early Jersey Wakefield' (dark green, compact, pointy head)

Carrot: 'Touchon' (long, deep orange roots)

Chicory: 'Red Treviso' (burgundy and white leaves), 'Red Verona' (bright red heads)

Corn: 'Golden Bantam' (high yields, 2 ears per stalk)

Cucumber: 'De Bourbonne' (small, great for sour pickles), 'Lemon' (small, round, yellow fruit)

Eggplant: 'Rosa Bianca' (rosy lavender fruit)

Lettuce: 'Bibb' (loose heads, dark green leaves), 'Black Seeded Simpson' (large, loose, light green leaves), 'Brune D'Hiver' (bronze-red leaves), 'Rouge D'Hiver' (crispy, red leaves), 'Tom Thumb' (tiny butterhead)

Melon: 'Charentais' (smooth rind, superb flavor)

Pea: 'Lincoln' (very productive, easy to shell), 'Petit Provencal' (very early, very productive)

Pepper: 'Habanero' (very hot, orange), 'Red Cornos' and 'Yellow Cornos' (sweet, ripen red or yellow)

Pumpkin: 'Rouge D'Etampes' (dark orange fruits)

Squash: 'Ronde de Nice' (zucchini with dark green, globe-shaped fruits), 'Zucchetta Rampicante' (S-shaped summer squash)

Swiss chard: 'Argentata' (deep green leaves)

Tomato: 'Big Rainbow' (very large, golden fruit), 'Brandywine' (large, purple-red fruit), 'Marmande' (meaty, scarlet-red fruit)

Turnip: 'Gilfeather' (white, sweet)

CULTIVATING AND PLANTING

Growing the best crops starts with learning how to get the most from your soil. A healthy soil yields healthy vegetables, and the techniques in this chapter will ensure your success. First of all, you'll discover the best ways to improve your garden soil. You'll learn how to identify the chemical and physical characteristics of your soil, the importance of good drainage, and the benefits of cover crops.

Composting is a very important aspect of organic gardening. It helps you to manage yard and kitchen wastes. When the finished compost is returned to your garden, it becomes one of your most valuable sources of organic matter and nutrients. In this chapter you'll learn different composting techniques and how to use the finished compost.

Once you have an understanding of these basic methods, you will be ready to get your vegetable garden under way. You will learn how to plant seeds indoors for an early start to the season. You will also be guided through all the stages until your seedlings are transplanted outdoors.

But before you transplant your vegetables outdoors, there are a few simple approaches you will need to consider. You will learn the benefits of crop rotation, companion planting, succession planting, successive sowing, and interplanting. All these techniques will benefit and increase your vegetable yields.

There are also advantages to be gained from raised vegetable beds, especially if you have poorly drained soil. You will gain an understanding of how to build and maintain them. And when all these facts have been taken into consideration, you'll learn how to extend your growing season with the use of cold frames and row and plant covers.

Opposite: Good soil preparation is always the key to a successful organic vegetable garden. A rich, loose, well-drained garden soil will encourage your healthy seedlings to grow into strong, high-yielding plants.

Soil Fertility and Productivity

Organic gardeners know there's a lot more than just chemistry behind large yields. Soil fertility and plant productivity are the result of high biological activity, good organic-matter content, and good soil structure. Make it a habit to monitor your soil's health, and make the necessary changes.

The best way to improve your soil's natural fertility is to increase its organic-matter content. This, in turn, will improve the soil structure and biological activity. It's also important to add organic matter to your soil when you till. Turning the soil adds huge amounts of oxygen, which increases microbial activity. This expanding microbial population needs food in the form of organic matter. Without it, the microorganisms will consume and deplete your soil's organic reserves. Remember also not to work the soil when it's either too wet or too dry. Working the soil in either of these conditions can lead to a damaged soil structure, which is difficult to repair. Read "Soil" on page 22 to learn more about the physical, biological, and chemical properties of soil.

Nutrient Analysis

Once you've chosen your garden site, have the soil tested for nutrient content and pH. While you wait for the results of the nutrient analysis, it's still possible to grow vegetables if you add organic matter and compost to the soil, and feed the growing plants with a soluble organic food like seaweed or fish emulsion. You can purchase soil-test kits from your local Cooperative Extension Service or from private soil-testing laboratories, or you can have a lab test the soil for you. Repeat the testing every 3 to 5 years, especially if yields decline or you suspect a nutritional deficiency. You may need expert advice to interpret the signs of nutritional deficiency, since symptoms are easily confused with insect, disease, or environmental pollution. Contact your local Extension Service for information and advice.

Testing Your Soil Collect samples at any time of year, but sample at the same time each year. You'll need a shovel or hand trowel, and a bucket; stainless steel or plastic tools and containers are best. Scrape away surface litter from the soil, and dig a small hole about 6 inches (15 cm) deep. Take a subsample by slicing away soil from the side of the hole. Collect at least 20 subsamples, regardless of garden size. In the bucket, mix the soil well before removing a 1-pint (600 ml) sample. Keep lawn and garden samples separate.

If you're having your soil tested by a lab, follow their specific instructions for collecting and handling samples. Most laboratories test for calcium, magnesium, sodium, sulfur, phosphorus, potassium, and trace minerals like copper and zinc. You can ask for special tests like organic-matter content, soil texture, and nitrogen. Most labs don't routinely test for nitrogen since it is so unstable in soil. Instead, nitrogen recommendations are based on the site history, and the crops you plan to grow. Be sure to request recommendations for organic amendments, because many laboratories routinely offer recommendations for only synthetic chemical fertilizers.

The amount of sand (left), and clay (right) in your soil will affect its productivity. The best type of soil, loam (center), is a mixture of both.

Soil Structure and Texture

Soil structure and texture are important because they influence all of the physical and chemical reactions that occur in soil, indirectly determining whether your vegetables grow vigorously or struggle to survive.

Soil structure refers to the way your soil's particles clump together. A soil with a poor structure may be very loose, so water and nutrients drain out very quickly, or quite solid, so that it forms dense clods when you try to work it. A good soil will have a loose but crumb-like or granular structure.

Soil texture refers to the relative proportions of sand, silt, and clay particles in the soil. Some soil-testing labs

Straw is a perfect mulch to improve your soil. Spread it on the soil and brush it aside when ready to plant.

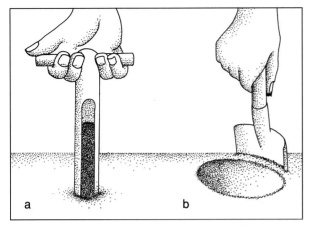

To take a soil sample, (a) use a sampling probe to remove a core of soil; or (b) dig a hole 6 inches (15 cm) deep, and slice a subsample from the side of the hole.

provide you with an analysis of your soil texture when you submit a sample for nutrient analysis. You could also consult a soil map of your area. These are available from local soil conservation agencies governed by the United States Department of Agriculture. A soil map tells you what soil types are most common in your area.

It's extremely difficult to change your soil's texture. However, you can improve soil structure by adding organic matter. To learn more about how structure and texture affect plant growth, see "Soil Structure" and "Soil Texture" on page 22.

Testing Soil Structure and Texture at Home Brush away surface debris and take a handful of fairly moist soil. If the soil is massed together in clods, or if it is so loose that you can see separate particles, your soil has a poor structure. A crumb-like or granular appearance indicates good structure.

Now squeeze the soil between your thumb and index finger to determine the texture. A gritty feeling is due to sand, just like the sand that runs through an hourglass or through your fingers at the beach. A slippery, plastic feeling is due to the presence of clay. Soil high in clay can be molded like cookie dough into ribbons and balls. If your soil feels greasy or like moist talcum powder, it contains a high proportion of silt.

Soils contain some proportion of all three minerals; it's likely that you will feel grit and smoothness in the same sample. Sections of your garden or yard may have different textures, so you should collect and examine samples from several locations. With a little experience, you can estimate the relative percentages of each mineral in your soil.

Soil Drainage

Soils vary enormously all over the country. The best gardening soils hold both oxygen and water in the right proportion within the soil pore spaces. If your soil is continuously wet, the drainage is inadequate.

Testing Soil Drainage at Home Watch for slow or excessive drainage after heavy rains. Avoid gardening at locations where rainwater forms puddles, or soil remains mucky for several days after a storm. When roots are flooded, plant stress symptoms include wilting, pale color, root decay, leaf dropping, and lack of vigor. All of these symptoms may also be caused by other disorders, such as pest or disease attack, so check the soil for waterlogging before you spray your plants for insects or disease. If soil is particularly sandy, water will drain too quickly and your plants will droop and wilt often. You can improve a heavy clay soil by double-digging or by adding organic matter or gypsum to it. If there are low areas of your yard that drain poorly, you can manage them with raised beds. Read "Raised Beds" on page 58 to learn how to build these structures.

The best source of organic nutrients is compost. Add at least 1 to 2 inches (2.5 to 5 cm) to your soil each year.

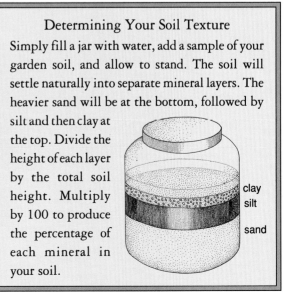

Determining Your Soil Texture
Simply fill a jar with water, add a sample of your garden soil, and allow to stand. The soil will settle naturally into separate mineral layers. The heavier sand will be at the bottom, followed by silt and then clay at the top. Divide the height of each layer by the total soil height. Multiply by 100 to produce the percentage of each mineral in your soil.

clay
silt
sand

Adding Organic Nutrients

The best thing you can do to improve or maintain garden yield is to add plenty of organic matter. This loosens up clay soils and improves the water retention of sandy soils. Organic materials can also supply the nutritional requirements of most vegetable crops.

The best source of organic matter is compost. Read "Composting" on page 46 to learn how to make it. If you have neither time nor resources to make your own, you can add the raw materials (like grass clippings, organic kitchen wastes, shredded leaves) directly to soil. Just allow more time for decay and nutrient release. It's a good idea to wait several weeks after adding fresh materials before planting. Spread the materials evenly over the soil surface, using a garden fork or shovel.

In most climates, you'll want to add 1 to 2 inches (2.5 to 5 cm) of compost each year, or 4 inches (10 cm) of fluffy material like straw. In warm climates, higher temperatures stimulate microbial activity and decomposition of organic materials, so you'll need to add more. No matter what the climate, it's difficult to add too much. Another way to add organic matter is by growing green manures. See "Green Manures" below.

You can spread organic materials on the soil before tilling and planting or leave the materials on the soil surface as a permanent mulch. Simply brush aside the mulch when it's time to plant. Be sure to add more each year. For more details, see "Mulching" on page 64.

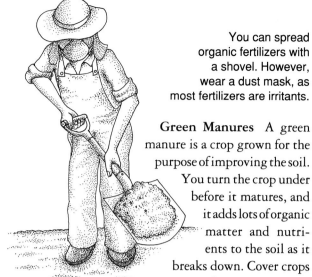

You can spread organic fertilizers with a shovel. However, wear a dust mask, as most fertilizers are irritants.

Green Manures A green manure is a crop grown for the purpose of improving the soil. You turn the crop under before it matures, and it adds lots of organic matter and nutrients to the soil as it breaks down. Cover crops are similar to green manures, but their primary purpose is to cover the soil and protect it from erosion.

Common green manures or cover crops include legumes like hairy vetch and red clover, and grasses and other crops like winter rye and buckwheat. Legumes are particularly good for building soil fertility by working with beneficial soilborne bacteria to transform nitrogen from the air into a form that plants can use. Different bacteria work with different legume crops. To make sure your crop has the right bacteria, you can buy them in a granulated mix (called an inoculant) and dust it onto your seeds before planting. For more information, see "Growing Green Manures" on page 45.

Organic Fertilizers

Use the results of your soil test to determine how much fertilizer to apply. If your soil is of low fertility, use the higher rate of application. If your soil is fairly fertile, you can use the lower rate or substitute a 1-inch (2.5 cm) layer of compost.

Fertilizer	Percent Nutrient			Application Rates
	Nitrogen (N)	Phosphate (P_2O_5)	Potash (K_2O)	lb/1,000 sq ft (kg/10 sq m)
Sources of Nitrogen				
Bat guano	10	3	1	10–30 (0.5–1.5)
Bloodmeal	11	0	0	10–30 (0.5–1.5)
Fish meal	5	3	3	10–30 (0.5–1.5)
Soybean meal	7	0.5	2.3	10–50 (0.5–2.5)
Sources of Phosphorus				
Bonemeal	1	11	0	10–30 (0.5–1.5)
Colloidal phosphate	0	2	2	10–60 (0.5–3)
Rock phosphate	0	3	0	10–60 (0.5–3)
Sources of Potassium				
Granite dust	0	0	4	25–100 (1.25–5)
Greensand	0	0	7	25–100 (1.25–5)
Ground kelp	1.5	0.5	2.5	5–20 (0.5–1)

Clover can be grown as a green manure. Turn the crop under before it matures, to add organic nutrients to the soil.

Adding Organic Fertilizers

Even heavy applications of organic matter, however, won't solve a major nutritional imbalance. In such cases, organic fertilizers made from natural plant and animal materials may help. The soil-testing lab will indicate if your soil has tested low, medium, or high for each nutrient. If the results indicate an imbalance of one or more nutrients, your first priority will be correcting it. Follow the lab's recommendations for applying organic fertilizers, or check "Organic Fertilizers" on page 44 to determine how much to use.

If you purchased a commercial organic fertilizer, look at the label to determine how much nitrogen, phosphorous, and potassium the product supplies. You'll find a fairly prominent three-part number that indicates the percentage of nitrogen (N), phosphate (P), and potash (K). For example, the analysis 1–2–3 indicates a fertilizer that contains 1 percent nitrogen, 2 percent phosphate, and 3 percent potash. The label should also list the appropriate application rate for that product.

You can spread fertilizers in your garden by hand or with a mechanical spreader or a shovel. Wear a dust mask, since most fertilizers are physical irritants if inhaled. Mark off the area to be covered and measure the right amount of fertilizer so you don't use more than is needed. If possible, mix or till in most organic

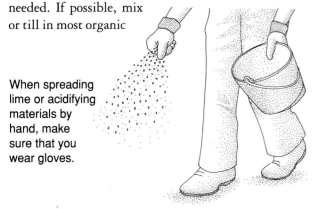

When spreading lime or acidifying materials by hand, make sure that you wear gloves.

Growing Green Manures

To plant any green manure crop, broadcast the seeds by hand or use a mechanical spreader, then lightly till or rake them into the soil. Eight to 12 weeks before you plan to plant your vegetable garden, till the crop under. If growth is heavy you might have to mow first.

- **Buckwheat:** Sow in vacant garden space between spring and fall crops. Annual; frost sensitive; tolerates low fertility; rate: 2 to 3 pounds/1,000 square feet (1 to 1.5 kg/93 sq m).
- **Rye:** Sow in late summer or fall, between fall and spring crops. It survives cold winters and resumes growth in spring. Winter annual; tolerates low fertility; rate: 2 to 3 pounds /1,000 square feet (1 to 1.5 kg/93 sq m).
- **Hairy Vetch:** Sow like rye, but add the proper inoculant (check with your seed dealer). Annual legume; winter hardy; tolerates most soils; rate: 1 to 2 pounds/ 1,000 square feet (0.5 to 1 kg/93 sq m).
- **Red Clover:** Sow in spring or summer, then allow to grow for 2 years. Biennial legume; tolerates acid soils with poor drainage; rate: 4 to 8 ounces/1,000 square feet (100 to 200 g/93 sq m).

fertilizers 6 to 12 months before planting. Nitrogen is the exception. Add nitrogen fertilizers like bloodmeal or aged manure just before planting. You can also add fertilizers at the same time you apply compost and other organic materials.

Adjusting Soil pH

Spread lime or acidifying materials by hand (be sure to wear gloves), or with a mechanical spreader or a shovel. The most common liming materials are calcitic limestone (calcium carbonate) and dolomitic limestone. Dolomitic limestone contains magnesium, so use it only if your soil-test results indicate a magnesium deficiency. Common acidifying materials include sulfur and peat moss.

The amount of material you must add to adjust soil pH depends on both its initial pH and the soil texture. The best course of action to take is to have your soil tested and follow the recommendations of the lab.

Composting

Composting is an excellent way to manage yard and kitchen wastes and make your own fertilizer. Compost is a good source of organic matter and nutrients for plants. It improves soil structure and water retention. And research shows that compost contains beneficial microorganisms that suppress plant pathogens in soil.

How to Make Fast, Hot Compost

1. Start saving all the organic wastes you normally throw away. Save grass clippings, leaves, kitchen scraps, rotted straw, and other organic wastes. Decomposition is most efficient if materials are shredded. You can use a leaf-shredding machine, but a lawn mower will shred leaves and other light materials to the right size for mulching. Things like woody prunings, tree bark, and newspaper should be shredded to fine pieces.

A few things you'll want to avoid adding to your pile are oils, meat scraps, and bones. They'll attract scavenging animals and slow the decomposition process. Other things to avoid include human or pet feces, and pesticides or pesticide-treated grass clippings.

2. You can designate a spot near your vegetable garden for composting, and store ingredients in piles or bins. While containment isn't important for composting, walls will keep the area neat, and Fido out. You can make small bins (about 3 feet [90 cm] in diameter) from chicken wire, or construct a permanent bin from untreated wood. You can also purchase a commercial ready-made compost bin, fence, or tumbler.

3. You'll need enough raw materials to make a pile roughly 3 feet (90 cm) on each side; smaller piles won't heat as efficiently. Build the pile by stacking the materials layer upon layer, using any kind of garden fork. Alternate the materials that are high in carbon (brown, woody materials like sawdust, straw, and newspaper) with layers of material high in nitrogen (green, sloppy materials like fresh grass clippings, kitchen scraps, weeds, and manure). You can achieve the right proportions with approximately equal volumes of high-carbon and high-nitrogen

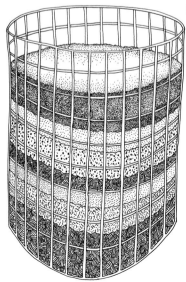

A compost bin enclosed in wire fencing is simple to make. The bin should be roughly 3 feet (90 cm) high for the compost to heat efficiently.

materials. A pile with too much carbon remains cool and breaks down more slowly. Too much nitrogen can create odor problems.

As you work, add several shovelsful of garden soil to inoculate the pile with the right decomposer organisms. Just sprinkle soil on top of alternate layers. You can add dry mineral fertilizers, too, if your soil needs them. If you are adding diseased or insect-infested plant material to your pile, make sure you put it in the center, where the high temperatures will kill the pests.

4. Keep the pile moist, but not soggy. As you build the pile, you can sprinkle the layers with water if the materials are dry. Keep the layered pile covered with a tarp to help maintain the right amount of moisture. Conditions that are too wet or too dry will change the rate of decomposition, and your compost might not be finished when you want it.

5. The organisms that perform decomposition require oxygen while they work. The simplest way to aerate the pile is to turn it every day or two. Using your garden fork, invert the pile one forkful at a time just next to the original pile. Materials originally on the outside should end up in the middle. Fluff the pile as you go, sprinkling with more water as needed.

Aeration hastens microbial activity, which increases the temperature of the pile. Turning every day or two helps maintain a constant temperature. You can use a compost thermometer to monitor your pile's temperature. The temperature should stay below 160°F (71°C), since higher temperatures will kill important decomposer organisms. Turning the pile will raise the temperature if it gets too cool. If the temperature gets too high, let the pile stand for a few days without turning it, or add water to it.

A hot compost pile can be finished in 2 to 6 weeks. You'll know it's finished when the temperature stabilizes and the individual materials you added at the beginning are no longer recognizable.

Finished compost provides all the organic matter and nutrients required by your vegetable garden.

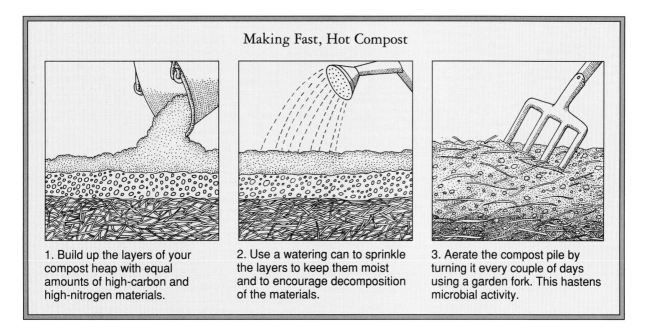

Making Fast, Hot Compost

1. Build up the layers of your compost heap with equal amounts of high-carbon and high-nitrogen materials.

2. Use a watering can to sprinkle the layers to keep them moist and to encourage decomposition of the materials.

3. Aerate the compost pile by turning it every couple of days using a garden fork. This hastens microbial activity.

How to Make Slow, Cold Compost

If you are willing to wait several months for your finished product, consider making a cold compost pile. (The pile won't actually feel cold; it just won't get as warm as a hot compost pile.) Simply pile all of your organic materials into a heap or bin (following the guidelines given in step 1 under "How to Make Fast, Hot Compost" on page 46), and let them sit for 6 to 12 months. You don't need to be as concerned about the balance of high-carbon and high-nitrogen materials, or about keeping the pile evenly moist. You don't even have to turn it, although turning will help speed up the process a bit without causing the pile to heat up. You will want to avoid adding diseased or insect-infected plant materials to the pile, since the pile won't heat up enough to kill the pests.

How to Use Compost

Use compost indoors as a potting medium for starting vegetable plants. First, shake the finished compost through ½-inch (12 mm) wire mesh to screen out large particles. The screened compost can be used alone, or mixed with vermiculite, perlite, or other potting ingredients. See "Growing Medium" on page 48 for more information about potting ingredients.

Outdoors, spread 1 to 2 inches (2.5 to 5 cm) of compost over the surface of your garden each year. If soil tests indicate a nutrient imbalance, sprinkle the appropriate fertilizers over the soil first, then add the compost, if you haven't already added fertilizer to the compost itself. You can dig the compost in, or leave it on top as a mulch.

Compost samples can be analyzed for nutrients, just like soil, so you know exactly what you are applying.

Generally, though, most homemade compost contains less than 2 percent each of nitrogen, phosphate, and potash, plus small quantities of trace elements.

Add water to your compost to make a liquid fertilizer suitable for seedlings and mature plants all season. Place about a quart (1 l) of compost in a cloth bag. Let it steep in about a gallon (4.5 l) of water for several days. For seedlings, dilute the solution with extra water. Use it full-strength on older vegetable plants in the garden.

Compost may be added as a side dressing to growing vegetables. You should apply the compost to a depth of about 1 to 2 inches (2.5 to 5 cm).

1. Choose a container that has drainage holes, and a light, moist medium like sifted compost. Fill the container with the growing medium and then press it firmly into the corners using your fingertips.

2. Level off the surface with a flat piece of wood, so that the medium is ½ inch (12 mm) below the rim of the container. If the medium is too high, it may spill over the edge of the container when you water.

Starting Seeds Indoors

Starting seeds indoors enables you to begin the new season earlier than if you plant outdoors. You'll have more control over the environment, so seeds will germinate faster and seedlings will grow vigorously.

Timing

Mark on your calendar the date of the last expected spring frost in your area. You'll have to count weeks *backward* from that date to know when to sow seeds indoors. Check individual plant entries in the "Plant by Plant Guide," starting on page 78, for specific times to sow, and when to move the plants outdoors. Slow-growing seedlings like leeks and onions will need a head start of 12 to 14 weeks. Several weeks later, you can sow peppers and cool-season vegetables like lettuce and broccoli. About 6 to 8 weeks before frost, start heat-loving plants like tomatoes and eggplant. Sow quick-growing cucumbers and squash only 2 to 4 weeks before the last frost.

Containers

You can start seedlings in just about any container that has holes for drainage. Seedlings started in their own pots won't require potting-up later. You can even purchase ready-to-use flats with cells for sowing single seeds. If you use containers such as peat pots or pellets, you can plant the pot along with the plant. They are a good choice for vegetables that transplant poorly, like cucumbers.

If you're reusing last year's flats and pots, either dip them in boiling water for several minutes or rinse them in a 10 percent bleach solution (one part household bleach to nine parts water) to kill disease organisms.

Growing Medium

New seedlings need a light, moist medium for a quick start. Fill containers with a medium like moistened vermiculite, milled sphagnum moss, perlite, or sifted compost. Materials like perlite and vermiculite don't provide any nutrients to your seedlings. Once true leaves have developed, the young plants will require extra nutrients. Water them with a liquid organic fertilizer like fish emulsion, at half strength. Gradually increase the dose to full strength. Alternatively, you can transfer seedlings to a potting blend with extra compost, or prepare a nourishing substitute by adding ½ cup (4 fl oz/125 g) or less of dry, organic fertilizer to each 5-gallon (22.5 l) batch of homemade potting medium. Remember, the longer the seedlings remain in pots, the more nutrients they'll need.

Sowing

First, prepare markers for each of the vegetables or cultivars that you plan to sow. Moisten the planting medium with warm water before filling the pots and flats. In pots or cells, plant two to four seeds in the center, leaving about 1 inch (2.5 cm) between the largest seeds. In flats, mark shallow furrows 2 inches (5 cm) apart, then sprinkle seeds into the furrows. Space medium-sized seeds like tomatoes ½ to 1 inch (12 to 25 mm) apart. Sow small-seeded vegetables like lettuce at ½-inch (12 mm) intervals. You should sow more thickly if you expect poor germination. Label each row of seed as you sow.

Lightly press seeds that need light for germination into the surface. Shallowly cover vegetable seeds that germinate in the dark with ⅛ to ¼ inch (3 to 6 mm) of fine planting medium. Mist the surface lightly with

5. To cover the seeds, sieve a light layer of growing medium over the surface of the soil. Do not cover seeds that need light for germination.

6. Water the seeds using a fine mist of water. You should keep the planting medium moist but not soggy. A spray mist bottle will help create the ideal conditions for germination to take place.

3. Mark out furrows, 2 inches (5 cm) apart, or just sprinkle the seeds evenly over the surface. Space medium-sized seeds like tomatoes 1 inch (25 mm) apart and small seeds like lettuce at ½-inch (12 mm) intervals.

4. Firm the surface by lightly pressing the seeds with a smooth wooden block or the palm of your hand. Do this gently to ensure the seeds are actually in contact with the mix, but not so hard as to completely bury them.

water. Seeds will rot if they are submerged in water.

Keep the planting medium moist but not soggy, and check daily for signs of the first sprouts. You can cover the containers with clear plastic (to permit light but retain moisture) or a layer of wet newspaper (to avoid light but retain moisture) before the seeds germinate, but remove any covering as soon as the seedlings appear. Ventilation is important as poor ventilation encourages the development of seedling diseases.

Temperature and Light

Most vegetable seeds germinate best at temperatures between 75 and 90°F (24 and 32°C), so keep your flats and pots in a warm place. On top of the water heater or refrigerator are both good positions. You can even buy horticultural heating mats to keep the soil warm. Place the mats on a heatproof surface and set flats or pots on top. Once the seedlings are up and growing, they will require less warmth and will do best at 60 to 70°F (15.5 to 21°C) during the day, with a 10°F (5°C) drop each night. Lower temperatures might injure cold-sensitive vegetables, and higher temperatures may tend to cause spindly, weak growth in your seedlings.

Once your seeds have germinated, it is important that you keep them near a sunny window or under fluorescent plant lights. The young plants will need between 10 and 16 hours of light each day. They will tolerate low-light conditions best if you keep temperatures on the low side.

Thinning and Transplanting

The first leaves to appear on the plant are actually specialized food-storage organs. These drop away quickly, and the leaves that follow are the first "true leaves." When seedlings develop their first set of true leaves, they'll need some special attention.

If seedlings are in separate pots or cells, thin away the extra seedlings by clipping them off at soil level with sharp scissors. One healthy specimen should remain in each pot or cell.

If your seedlings are crowded together in a flat, you'll need to transfer them to their own containers. Fill pots or cells with a nutrient-rich medium. (For information on choosing a medium, read "Growing Medium" on page 48.) Scoop one or more seedlings from the flat (a teaspoon works well), separating their root systems as gently as possible with your fingers. Poke a hole in the potting soil deep enough to accommodate the new seedling and its root system, then transfer it and firm the soil. Water lightly, and place pots away from direct light and wind for a day until the shock of transplanting is over.

Hardening Off

Hardening off allows young plants to adjust slowly to the extremes of wind, temperature, and light. At least two weeks before transplanting time, begin watering less frequently and withhold fertilizer. One week before you transplant, move the plants outdoors to a spot protected from strong light and wind. Starting with 1 or several hours, gradually increase their time spent outdoors. Within a week, they should be outdoors permanently. They'll need more water at this time, since sun and wind quickly dry the soil. Be prepared to bring them back indoors if a cold snap or storm threatens. Refer to "Transplanting" on page 50 for information on transplanting hardened-off seedlings to the garden.

7. Prepare markers for each species or cultivar, and label all containers with the name and the date of sowing. It's easy to forget which plants are which, as many seedlings look similar.

8. Cover the container with glass or plastic wrap to keep the seeds moist and warm. Check daily for signs of the first sprouts and remove the covering as soon as the seedlings appear, to prevent disease.

Mark out rows using a length of string as a guide to make sure your rows are straight. The seed packet will indicate how far apart seeds should be spaced.

Planting Outdoors

Some plants like beans, peas, and root crops, do not transplant well. You'll get the best results by planting these seeds directly into the garden. Planting seeds outdoors is not a good idea for seeds that are hard to germinate, or for vegetables that need a long, warm growing season.

Sowing Seed

1. Before the season begins, check your seed packages or catalog or the "Plant by Plant Guide," starting on page 78, for the best time to plant outdoors. Planting dates vary among vegetable species and cultivars. You'll need to know the expected date of the last frost for your area; ask local gardeners, or contact your local Cooperative Extension Service.

2. If you're planting in rows, mark them before planting with wooden stakes at both ends. Stretch a section of lightweight string down the row, and use it as a guide to keep your rows straight while you plant. Transfer it from row to row as you plant.

3. If you garden in raised beds, divide the beds up into sections by plant cultivar, broadcast the seed over the appropriate section, and label them. You can also simply sow your seed in rows across the raised bed.

4. Follow the spacing guidelines on the seed packet, or check the "Plant by Plant Guide" to space seeds within the row or bed. Some seed packages indicate how many feet of row the package will sow. Most sowing recommendations are higher than necessary, to account for environments that are less than ideal. If conditions are good for germination, you may have to thin your seedlings later.

5. Cover the seeds with soil, to a depth two to three times the seed diameter, then firm them by pressing the soil lightly with the flat side of a rake or hoe.

6. The soil should remain moist but not soggy until your seeds begin to sprout. If you plant during a dry season, you may have to water to ensure germination. If the soil at the planting depth is moist when you sow large seeds, they should have sufficient moisture for germination. Fine seeds may need extra moisture. You can cover the row with lengths of board to retain moisture. Check the soil under the boards frequently, and remove them as soon as seeds sprout.

7. Once seedlings are growing vigorously, you can thin them to the final plant spacing; check the seed package or the "Plant by Plant Guide" for the recommended spacing. Pull or snip off extra seedlings, leaving only the healthiest.

8. As you work, keep records of your gardening activities, and make a map to record the location of each vegetable species and cultivar.

Transplanting

It's best to transplant on a cool, overcast day. Give your seedlings the best start by working in the evening, so they'll have the night to recover. You should avoid

1. Seedlings are best transplanted on cool, overcast days and in the early evening to shield them from the sun.

2. Gently slide each seedling out of its container, making sure that you leave the soil and rootball intact.

transplanting during hot, sunny, or windy weather.

1. Check the seed package or the "Plant by Plant Guide" for the best time to move transplants to the garden. If you started your own plants from seed, harden them off before transplanting. To learn how to do this, read "Hardening Off" on page 49.

2. Mark rows or beds as you would for planting seeds. Follow spacing recommendations for transplants, and plant only the healthiest specimens. Once you've filled your rows or beds, don't toss away the extra seedlings—save them for a week or two to fill empty spots in case a few of your transplants don't survive.

3. Using a hand trowel, dig a hole slightly wider than the root ball. Tap the plant from the container and set it in the hole, slightly deeper than it was in the pot. Replace the soil over the roots, and gently firm it. Some vegetables, like tomatoes and cabbage, grow excessively tall as seedlings and need to be transplanted deeply. Perennial vegetables, like asparagus, need special attention since they'll grow in the same location for 20 years or more. Check the individual plant entries in the "Plant by Plant Guide" for special instructions on transplanting depth.

Seedlings in peat pots also need special handling. Tear a hole in each side of the pot before you plant it,

3. Holding the stem gently, plant each seedling in a hole slightly wider but of the same depth as the container.

4. Replace the soil around the roots first, fill each hole level with the bed, and gently firm the surface.

When young plants are well established, they may need to be thinned. Pull them out carefully, or snip them off at soil level.

Before transplanting seedlings grown in peat pots, tear off the rim of the peat pot below the soil level in the pot. Then put slits in the pot's sides and tear off the bottom, so the roots are able to get through.

so that roots can escape. Then, tear away the upper collar of the pot, since exposed peat acts like a wick to draw moisture away from the roots, and plant as you would a normal seedling.

4. Water the soil around each plant. Use at least 1 quart (1 l) of water per plant.

Planting Bulbs

Bulb-producing vegetables, like onions, garlic, and shallots, are among the easiest vegetables to grow. You can plant garlic and shallots in the fall, and they'll survive a severely cold winter, sending up green shoots the following spring. Grow perennial onions, and you'll only have to plant once.

When you prepare garden space for bulbs, dig in plenty of bonemeal or rock phosphate, or add a teaspoonful to each planting hole. Bulbs need more phosphorous than leafy vegetables, and less nitrogen. Plant bulbs individually or in clumps. Poke a hole in the soil to the right depth, and position the bulb with the pointed end up. Cover with soil to a depth of 1 inch (2.5 cm). Check the "Plant by Plant Guide" for specific depth and spacing guidelines.

Marigolds are said to control root-feeding nematodes. Plant them thickly as a cover crop, and turn them into the soil at the end of the season.

Companion Planting

Companion planting means growing a *planned* diversity of vegetables, herbs, and ornamentals together. Pest control is the aim, but the benefits of companion planting include better space and nutrient efficiency.

Companion planting is based almost entirely on folklore passed through generations of observant gardeners. Although modern scientists studying the relationships among plants and insects recognize the advantages of companion planting, more investigative studies are necessary before recommendations can be given with confidence. Most examples, therefore, are based on folklore. Put companions to work in your garden by arranging them as borders or mix them with the vegetables in your rows and beds.

Insect Attractants and Repellents

Beneficial insects, like parasitic wasps or predaceous flies, depend on pests for nourishment at one or more stages of their life. Attractant plants are so named because they offer an alternative food source for beneficials when pests are in short supply. They also provide shelter from heat and wind, and protection from natural foes (even the good insects have enemies).

Many attractant food plants are members of the carrot and daisy families. They have lots of small flowers to provide the pollen and nectar that maintain beneficial insect populations. Encourage beneficial insects by mixing flowering plants among your vegetables. Good choices include angelica, anise, dill, fennel, Queen Anne's lace, sunflowers, yarrow, and zinnias. Read "Biological Controls" on page 73 for more information.

Other plants, like tansy and mint, will discourage pests. Repellent plants directly repel or kill pests, or help to disguise the attractive characteristics of vegetables (like scent or color) that pests search for. Scented marigolds, planted thickly and turned into the soil, ward off pests below ground. Many herbs, like basil, coriander, catnip, marjoram, mint, parsley, rue, santolina, tansy, and thyme, repel pests from vegetables planted nearby. Mix and match to see which combination works best for you.

Trap Crops

Some plants are so attractive to pests, you can use them to distract the pests away from your vegetables. Plant a border of nasturtiums, for example, to attract aphids and flea beetles away from your crops. Early plantings of squash will attract pickleworms from your melons. Other plants and the pests they attract include chervil

Cabbage, lobelia, and yellow calendulas make colorful companions. Calendulas also attract beneficial insects, like hover flies.

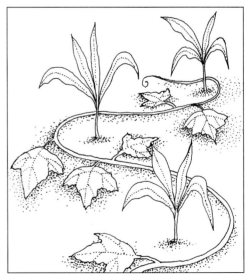

Shallow-rooted vines are good to plant at the base of deep-rooted corn stalks.

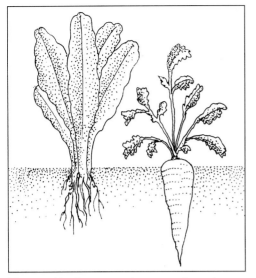

Root and leaf crops such as carrots and spinach are an effective combination.

Herbs are excellent companions for vegetables. Cabbage pests are repelled by mint, while basil repels tomato hornworms.

(earwigs), dill or lovage (tomato hornworm), and potatoes (wireworms). Watch for the build-up of pests on the trap crop, then either kill them with a spray of soap or botanical insecticide, or pull the crop and destroy it.

Compatible Plants

Some plant combinations are beneficial for environmental or nutritional reasons. If you mix together deep- and shallow-rooted vegetables, for example, you'll use garden space and nutrients more efficiently. A combined planting of climbing beans and corn is one of the oldest examples. Since beans are legumes, they fix nitrogen that ordinarily would not be available to the corn; and corn is a sturdy support for the bean vines. Tall corn plants also provide shade for summer plantings of lettuce. Still other plants, like basil and tarragon, are said to benefit anything that grows near them.

Antagonists

You can prevent some problems by avoiding certain combinations. Most important, plant related vegetables, like tomatoes and potatoes, as far apart as possible, since they attract the same pests. Consider environment when you mix plantings. For example, don't shade other vegetables with a plot of corn.

"Alleopathy" is the term scientists use to describe the way some plants repel other plants. Walnut trees, sunflowers, and rye are examples of plants that secrete toxic substances into soil, through their roots. As a result, other plants grow poorly or die when planted nearby. Weeds like quack grass and nutsedge also have this advantage over other plants, including your vegetables. Watch out for these detrimental combinations when trying to diagnose plant ailments.

Sunflowers are beautiful, but they may inhibit the growth of other plants. Give them their own spot in the garden.

Nasturtiums entice aphids away from the vegetables and deter whiteflies and squash bugs.

Interplanting

You'll get twice the harvest variety if you interplant your vegetables. Interplanting means planting two (or more) kinds of vegetables *at the same time, in the same space*. Mixing vegetables enhances pest prevention, since many insect pests can't locate their favorite plants when scents and sights are mixed. More foliage means a greater variety of hiding and feeding places for beneficial insects that help control pests. You'll save gardening space, too.

To interplant in your garden, determine the length of your growing season, then choose several early- and late-crop combinations to fit. The first vegetable should mature quickly, like radishes or lettuce. It should be easy to harvest, without damaging the remaining crop. (For example, potatoes aren't a good first crop if digging will destroy the interplanted crop.) Or you could plant a traditionally long-season crop like leeks and harvest it at an early, tender stage. The second vegetable should be slow-growing, like melons, for a harvest scheduled later in the season.

The best crops for interplanting are unrelated and have complementary nutritional needs. You can plant both crops at once, or spread planting dates apart as necessary. Fulfill space requirements of the late-maturing vegetable first (since it will be there all season), then fill in extra space with the quick crop. Plant both vegetables at a lesser density, since moisture and nutrient demands on the soil are greater. If necessary, supplement the second crop with liquid fertilizer or a midsummer application of compost.

Fast-growing lettuce and slower-growing cabbage-family crops make a good combination for interplanting.

Suggestions for early vegetables are: amaranth, arugula, baby beets, carrots, cress, garden peas, leeks, lettuce, mustard greens, oriental greens, radicchio, radish, scallions, spinach, summer squash, and turnip. Suggestions for late vegetables are: beets, brussels sprouts, cabbage, carrots, cauliflower, cucumbers, eggplant, leeks, lima beans, okra, onions, peanuts, peppers, pumpkins, snap beans, sweet corn, sweet potato, Swiss chard, tomatillo, tomatoes, and winter squash. Consult the "Plant by Plant Guide," starting on page 78, for more crop ideas.

Interplanting gives you a greater variety of vegetables at harvest time. It is also helpful in preventing pests.

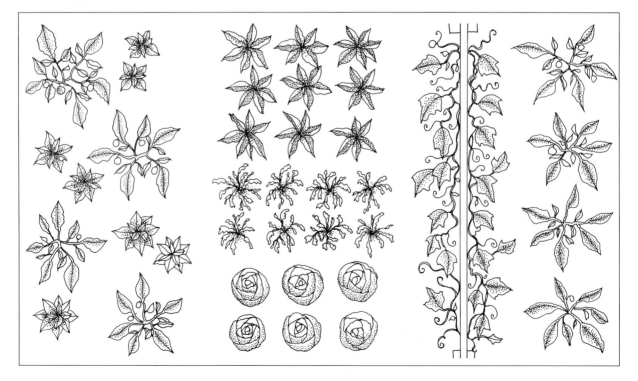

Successive sowings of crops like bush beans will ensure a continuous harvest. As one planting finishes bearing, the next is ready to harvest.

Succession Planting

Like interplanting, succession planting means harvesting two crops from the same space in one season. Succession planting, however, means you don't plant the second crop until after you've harvested the first.

Determine the length of your growing season by counting the number of days between the average date of the last spring frost and the average date of the first fall frost. Then select two vegetables whose combined days to maturity fit that limit. Plant seeds or transplants of the first crop in early spring, then harvest it and replant with the second crop in midsummer. Between crops, prepare the soil as needed but avoid tilling. By midsummer, most weed seeds have germinated, and tillage will only bring more seeds to the surface. You can add compost before the second planting, or after planting, as a mulch.

Timing is the challenge, since the first and second plantings don't overlap. Each crop must get off to a quick start and finish. Good choices for quick crops to grow in spring and fall are: arugula, beets, carrot, collards, chicory, endive, kohlrabi, mustard greens, radish, shallots, spinach, and Swiss chard.

Successive Sowing

Another way to ensure a continuous harvest is to make successive plantings. Sow the same carrot cultivar, for example, at several locations at 10- to 14-day intervals until the last harvest approaches the season's end. Lettuce, carrots, snap beans, and radicchio are good candidates for successive sowings.

Or reap successive harvests by planting several cultivars of the same vegetable, with a range of maturity dates, on the same date at the beginning of the season. For example, choose sweet corn cultivars that mature on a range of dates from 65 to 90 days. Familiarity with the cultivars will help in making your selections, and seed catalogs are the best reference to consult. You can make comparative charts and lists of cultivars once current seed catalogs begin arriving in early winter.

Chicory is an ideal quick-growing crop for succession planting. Grow it either in spring or fall.

Interplant carrots and onions. The carrots will be harvested before the onions have reached maturity.

Crop Rotation

You're already practicing simple crop rotation if you avoid planting the same vegetable in the same place each year. Most organic gardeners practice crop rotation for several important reasons. By rotating crops, you'll prevent the build-up of pests that might result if the same crop were grown continuously. You'll also make better use of soil nutrients, since different crops remove nutrients from the soil at different rates.

There are several steps you can take to get the full benefits of crop rotation.

1. On paper, divide your crops into different categories according to their growth habits or families. You might, for example, separate them into root crops, leafy crops, and fruiting crops. Or you might group them according to their respective families, such as Gramineae for corn, Cruciferae for cabbage and its relatives, Leguminosae for peas and beans, and Solanaceae for tomatoes and potatoes. If you don't know what family your crop belongs to, look next to its botanical name in the "Plant by Plant Guide," starting on page 78.

2. Draw a map of your garden, and divide the growing space into the same number of areas as you have categories of crops. If you are just using root, leafy, and fruiting crops, you'd divide your garden into three areas.

3. Plant one type of vegetable, or a group of vegetables, in each area of the garden. Record on your map where each is growing.

4. Next year, move each crop or grouping to a different area.

Grow a heavy-feeding crop like tomatoes after a light-feeding root crop like radishes.

Rotate potatoes (Solanaceae family) with vegetables from the Cruciferae or Leguminosae family.

Crop Rotation Hints

- Wait 3 to 5 years before growing the same vegetable, or a closely related one, in the same location.
- Include soil-improving crops like rye, buckwheat, or clover (see "Green Manures" on page 44 for more information on how to grow and use these crops). You can sow them between early and late crops, or between fall and spring crops.
- Grow legumes like clover or peas the year before grasses.
- Grow light feeders (like root crops) before heavy feeders (tomatoes, peppers, cabbages, and most leafy vegetables).
- Make heavy compost applications to heavy-feeding vegetables like corn and squash anytime during the growing season.
- Make light compost applications to light-feeding vegetables anytime during the growing season.

Legumes like peas improve the soil. Corn will thrive when planted where peas grew during the previous year.

Fruiting crops like peppers may be rotated with root crops like turnips or parsnips.

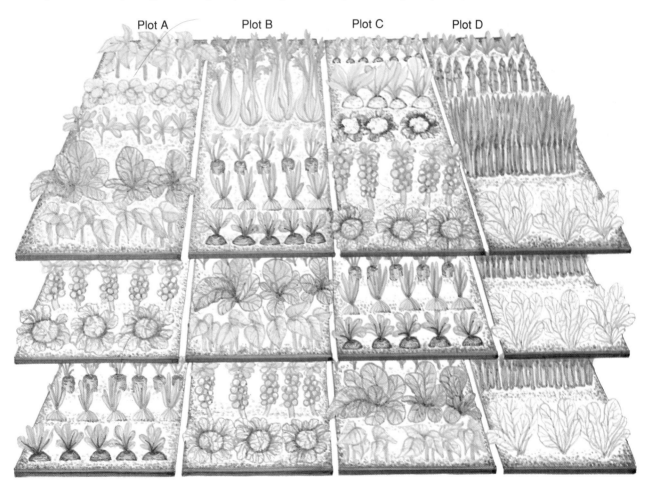

Plot A   Plot B   Plot C   Plot D

With raised beds, organizing crop rotations is fairly simple. Each year, move your crop over to the next bed in the sequence. Set one bed aside for perennial vegetables, like asparagus and rhubarb.

Raised Beds

Raised beds are the ultimate in outdoor, custom-built plant containers. They require an initial investment of energy, but you'll save both time and labor later.

Raised beds are more productive than rows for several reasons. Since traffic is restricted to paths, soil in dug beds remains loose and friable. Building a bed gives you the chance to correct both physical and chemical soil imperfections. If your soil drains poorly or lacks certain nutrients, this is an excellent time to make the necessary changes. Since the soil environment is ideal, you can plant closer than usual and harvest more. And, since plants are closer, the soil stays shaded and cool so weeds are less likely to develop, and less soil moisture is lost to evaporation. Since beds are discrete units, they're easier to manage when it's time to plan crop rotations. You can build beds wherever you like, in partial shade or full sun, depending on the site and your plants.

Raised beds save space, time, and money, and yield twice as many vegetables as ordinary garden rows.

Bed Maintenance

Each season, work more organic matter into the beds. Use a shovel or fork, or a rotary tiller if the beds are large enough. Include beds in your crop rotation, and allow them to rest with a cover crop every 3 to 5 years. When it's time to test the soil, you can lump the beds together as one if they've been treated identically. But if you've managed them individually, sample them individually. To avoid compaction and keep your beds in top shape, never walk on them.

Bed Size

Make your beds narrow enough so you can reach one half from either side. Four feet (1.2 m) is a good width for most people. If you find bending and reaching difficult, make them narrower. Beds can be as long, short, straight, or curved as you like. When building them on uneven ground, make them perpendicular to the slope to prevent soil erosion.

Leave 4 to 5 feet (1.2 to 1.5 m) between beds as paths, or as much room as necessary for yourself and your equipment. Paths can be paved, mulched, or planted. If you plant grass, make sure the paths are wide enough to accommodate your lawn mower.

Digging

There are at least two ways to form your beds: by simply incorporating lots of organic matter into the soil, or by double-digging. Before you start either method, decide where to locate beds and paths, and mark off the boundaries with flags or stakes and string. After digging, rake the tops smooth and allow the soil to settle for several weeks before planting.

If you're satisfied with your soil, the simplest method is to simply dig in plenty of organic matter like compost, rotted manure, or shredded leaves and straw. It's almost impossible to add too much organic matter. Using a shovel, work the top 6 inches (15 cm) of soil. Work one strip at a time, moving backward so you're always standing on undug ground. As you work, add other amendments (like lime or rock phosphate) as needed. When finished, you'll have a mound of soil.

If you soil is heavy and compacted or poorly drained, consider investing some more time and labor to double dig your raised bed. Double-digging will help to loosen up the soil and improve drainage. To use this method, mark a trench about 12 inches (30 cm) wide and as long as the width of the finished bed. Remove the soil to a depth of about 12 inches (30 cm) and place it in a wheelbarrow or on a tarp laid on the ground. Once you've reached the bottom of the trench, insert a

Leave enough room for yourself and your equipment between beds. Gravel is a popular surface for paths.

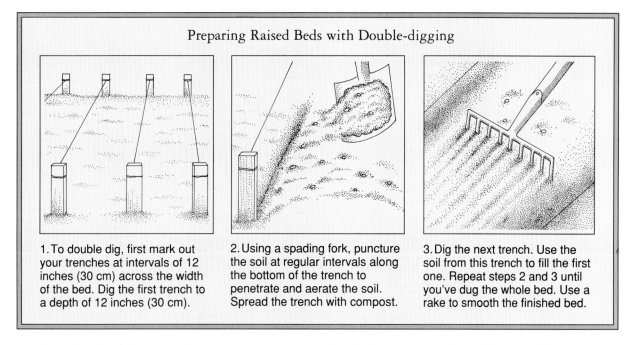

Preparing Raised Beds with Double-digging

1. To double dig, first mark out your trenches at intervals of 12 inches (30 cm) across the width of the bed. Dig the first trench to a depth of 12 inches (30 cm).

2. Using a spading fork, puncture the soil at regular intervals along the bottom of the trench to penetrate and aerate the soil. Spread the trench with compost.

3. Dig the next trench. Use the soil from this trench to fill the first one. Repeat steps 2 and 3 until you've dug the whole bed. Use a rake to smooth the finished bed.

spading fork and push on it with your foot to penetrate and aerate the soil; repeat at intervals of several inches along the length of the trench, then spread a layer of compost on the punctured surface.

Next, move 12 inches (30 cm) over on the soil surface and once again remove the soil to a depth of 12 inches (30 cm), but this time toss the soil into the first trench. When you've reached the bottom of the second trench, aerate with the fork and apply compost, then fill the second trench with soil from the third trench you have dug. Repeat this process until you've moved down the length of the new bed, filling the last trench with the soil saved from the first trench. As you work, avoid standing on the newly turned soil.

Framing

Raised beds can be left as mounds of soil, or you can add permanent sides. A frame allows you to build deeper beds, helps to prevent the soil from washing away, and defines the bed border. If the frame is sturdy and wide enough, you can stand or sit on it while you garden. Simply place the material you are using around the edge of the bed, and press it firmly into the soil. Wood, bricks, cement blocks, flat rocks, old construction timbers, and railroad ties are all good materials for frames.

Solving Drainage Problems

If your soil drains poorly, you can correct the problem by growing your crops in raised beds. Double-digging alone might solve your drainage problem. Or build a bed frame and import topsoil and organic matter to fill it, without mixing the native soil. If you're forced to garden in a low spot where water forms puddles, build a framed raised bed and put a layer of crushed stone in the bottom. Or, place a length of drainage pipe in the bottom, before filling with improved soil.

Naturally rot-resistant wood is always a popular choice for framing raised beds.

Crop rotation—moving crops each season to maintain soil fertility—is most easily managed in raised beds.

Use cold frames to harden off young seedlings and transplants and get them used to the cold and the sun.

Season Extension

Most gardeners would like to squeeze a little more time into the season, at the beginning or end. Season extension means getting a head start in the spring, and prolonging production in the fall.

Growing Frames

Use a growing frame for hardening off tender transplants and for growing cold-tolerant vegetables when garden temperatures are too low. The most simple growing structures are wooden, bottomless, box-like frames topped with a window sash roof. They're usually called cold frames, since they're unheated.

If you build your own, make the back taller than the front so the glass roof slants at an angle of about 45 degrees, for the best light. A frame measuring about 4 feet wide by 4 feet long (1.2 m by 1.2 m) is a good size. It should be tall enough so your plants can grow in their pots or flats without touching the roof. You can attach the roof at the back with hinges and prop it open with a piece of wood to allow for air circulation. Or the roof can simply sit on the frame and you can slide it open when you need to reduce the temperature inside the frame. Paint all wood surfaces to prolong the life of the frame, but first seal any leaks with caulking. The interior should be painted white, to reflect light onto your plants.

Place the frame facing south on a level, well-drained site near a source of water. You can make the frame permanent or portable, to store inside during the off-season.

Cold-tolerant vegetables, such as cabbage transplants, are ideal to house in growing frames in early spring.

Construct permanent frames on foundations that you have dug into the ground or built on the surface. Portable frames are simply bottomless boxes set directly on the ground.

It's best to keep potted plants or trays of seedlings in the frame instead of sowing directly into the soil floor. You'll find that containers are easier to manage in the small space. But if you choose to sow directly, prepare the soil the same as you would garden soil.

Most homemade cold frames are built to fit around discarded window sashing. Glass is inexpensive, and allows sunlight in while protecting plants from wind. You could also make a roof from fiberglass, Plexiglas, heavy polyethylene, fabric, or screening attached to a frame. Use heavy, clear materials to keep heat in during spring, fall, and winter. Use window screening or fabric (like spun-bonded row covers—see "Row and Plant Covers" on page 61) to promote ventilation and keep insects out during summer.

In early spring, use your growing frame to house cold-tolerant vegetables like lettuce or cabbage transplants. If cold nights are a threat, be ready to cover the closed frame with thick blankets or hay, or place bales of hay around it; remove the covers in the morning. Once the earliest plants have been moved to the garden, you can use the frame to harden off cold-sensitive plants. Use a thermometer to monitor the temperature inside, and prop open the roof on sunny days. Even on cool days, the combination of sunlight and moisture inside a closed frame can stress small plants.

After the threat of frost has passed, continue using the frame to hold vegetables until there is room to plant them in the garden. In the fall, move your cold-tolerant crops inside the frame. Prepare the soil just like in your garden, if you plan to sow seeds on the soil floor. In late summer, begin sowing carrots and spinach. You can harvest as long as the soil hasn't frozen.

Hot beds are simply growing frames with a heat

A plastic row cover will protect your plants from frost, trap warmth inside, and extend the growing season.

source; they are useful if you live in an area where it snows. Our great-grandparents used fresh horse manure, known for its natural heat, to warm growing frames in early spring. To use this method, dig a pit 6 to 12 inches (15 to 30 cm) deep in fall to fit your growing frame, and cover it with a tarp to keep snow out. Several weeks before you plan to set plants out in the frame next spring, fill the pit with fresh horse or cow manure and top with about 4 inches (10 cm) of soil, then replace the frame. Set potted plants or trays on top of the soil inside the frame, and cover as usual. Monitor the temperature daily and adjust the ventilation to provide the best environment.

Today, most gardeners rely on electric heating cables instead of fresh manure to keep their seedlings warm. First, dig a pit 12 inches (30 cm) deep to fit your frame. Follow the manufacturer's instructions for placing the cables correctly so they don't cross. Sandwich the cables at the bottom between layers of sand for drainage. Top cables with a screen or hardware cloth to protect them from damage, then cover with 4 to 6 inches (10 to 15 cm) of sand. Push pots into the sand to secure them.

Row and Plant Covers

You can start growing in the garden earlier, and prolong your fall harvest, if you use row covers. They are made from fabric or transparent plastic, and provide from 2 to 7°F (1 to 4°C) of frost protection at night. Lightweight row-cover fabric, made from spun-bonded polyester or polypropylene, allows water, light, and air in but keeps insects at bay. You can buy shaded or colored covers to reduce sunlight and heat in hot climates. Read "Barriers" on page 72 to find out how to use floating row covers.

Plastic row covers will keep plants and soil warmer than fabric. Since heat will build up on sunny days, you'll need to vent plastic covers. You can buy slitted plastic covers with built-in ventilation. To use, stretch plastic row covers over wire hoops inserted in soil on opposite sides of the bed or row. You can buy the supports and covering together.

If you garden on a smaller scale, you can protect individual plants (instead of rows) from frost. Many different types of material are used to make plant covers (also known as cloches). Whatever you use, make sure the covering allows for adequate light and ventilation. Garden-supply catalogs and stores offer many choices. Or make your own protectors from plastic milk jugs. Cut along three sides of the base of the jug, so the base forms a flap that can be used to anchor it. Leave the top open during the day. Remove the jug once the danger of frost has passed.

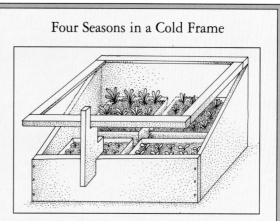

Spring: Harden off seedlings in your cold frame before transplanting to the garden.

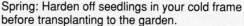
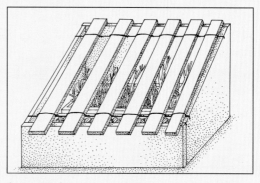

Summer: Sow salad crops like spinach for fall harvest. Lath shade keeps plants cool.

Fall: A cold frame ensures survival of late-fall leaf crops like lettuce well after the first frost.

Winter: Use your cold frame to give spring bulbs their cold treatment before forcing.

MAINTAINING YOUR GARDEN

Once your garden is prepared and planted, you'll need to spend some time taking care of those little plants to get them off to a good start. Regular maintenance—mulching, watering, and weeding—keeps your vegetables growing at a steady pace while preventing pest and disease problems from getting out of hand.

You'll find that simple measures are often the most effective in the garden. Mulching your plants with organic materials, for example, is one of the most essential aspects of organic vegetable growing, and it's central to good garden maintenance. Recycling plant residues onto the garden in the form of mulch or compost enriches the soil and provides nutrients to the plants. Mulching will also save you time, since a well-mulched garden needs watering and weeding less frequently. It can even help reduce pest problems.

When and how to water your garden are other critical considerations dealt with in this chapter. Understanding the needs of your crops will help you provide the right amount of water at the right time—before your plants wilt and their yields are reduced by drought stress. You'll also find that an awareness of the different watering methods available can save you time and money. Systems like drip irrigation and soaker (dew) hoses can help you apply water right where the plants need it, and eliminate waste due to evaporation or runoff, so you'll be making the most efficient use of this precious natural resource.

At some point in the season, you'll probably have to deal with weeds. You may be surprised to learn that some weeds can even benefit your crops, by indicating soil imbalances or providing food and shelter to beneficial insects. But serious weed problems can compete with your crops for light, water, and nutrients. A mulch is the first step in suppressing weed growth, but you'll also learn many other tips and tricks for keeping weed problems to a minimum.

For many people, space is a precious commodity in the garden. Fortunately, there are plant-training techniques that can help you get the most out of every bit of available space. You'll find out how to create useful staking and trellising systems for all kinds of crops, from beans and tomatoes to cucumbers and melons.

No matter how much care you take when planning your garden, there is always a chance that pests and diseases could cause a problem. Inspecting your vegetables regularly and protecting them with barriers like row covers are helpful preventive measures. If these aren't enough, you'll learn about safe and effective methods that you can use to keep plant pests and diseases under control.

Bountiful harvests will be your reward for all the care you take in planning and caring for your vegetable garden. And by following the guidelines discussed at the end of this chapter, you'll learn the best ways to harvest and store your vegetables in order to retain the best flavor and nutrition possible.

Opposite: When natural rainfall is lacking, regular watering is vital for a good harvest. Drought-stressed vegetable plants will produce less and they are also more prone to pest and disease problems than healthy plants.

Used as a mulch, strips of aluminum-coated paper or heavy-duty aluminum foil can confuse plant pests and prevent them from landing on your crops.

Mulching will boost garden productivity by enriching the soil (during decomposition) with plant nutrients, thereby encouraging earthworms and friendly bacteria.

Mulching

Many gardeners consider mulching to be their best investment of time since the returns are so great. Mulching controls weed growth, retains soil moisture, provides a barrier between soilborne diseases and your plants, protects soil from erosion, and keeps fruit and leaves clean. You'll stay clean, too, working among clean plants and walking on covered paths. Depending on the material you use, there are other benefits as well.

Organic Mulch

Organic mulches are plant residues, like compost, bean hulls, grass clippings, shredded leaves, newspapers, pine needles, sawdust, straw, and wood chips. In general, they enhance soil productivity. They enrich the soil with plant nutrients during decomposition, as well as supplying organic matter.

Organic mulches shelter and encourage the presence of beneficial organisms at the soil surface and just below it. Thick layers of mulch act as a cushion, reducing soil compaction. And at the end of the season, you don't have to remove them since they decompose naturally.

The vegetables you choose to grow and the location of your site will determine when and how you should apply mulch. Ideally, you should apply organic mulch before annual weeds (like lamb's-quarters) have a chance to germinate, and perennial weeds (like quack grass) emerge. Vegetables like peas and spinach will appreciate an early mulch, which can help keep the soil cool. Early mulching is also a good idea in dry climates, since it traps the moisture from spring rains. On the other hand, if you are planting heat-loving crops like eggplant or peppers, you want to wait until the soil has warmed up before you apply the mulch. Mulching later in the season is also useful in wet climates, since you'll give the soil a chance to dry out a bit.

You can build mulch layers 4 to 12 inches (10 to 30 cm) thick. Use less material in wet climates, and more in dry climates. Keep the mulch 4 to 8 inches (10 to 20 cm) away from plant stems—you'll lessen the chance of slug and snail problems, and promote surface aeration, which helps prevent disease. Use a light-colored mulch like straw to reflect light and keep soil cool. Dark-colored mulches will help warm the soil.

Use whatever organic materials are handy. You can purchase materials in bags, haul away free loads of shredded leaves from your community collection site, or make your own compost.

You'll have to amend some types of organic mulch. Pine needles and leaves, for example, are acidic and should be neutralized with lime. (As a general rule, applying 5 pounds [2.2 kg] of lime per 100 square feet [93 sq m] will raise the pH of your soil 1 point. Sandy soils generally need less lime to raise the pH; clay soils

Even shredded sheets of newspaper can be used as mulch. Wet the papers to keep them from blowing around.

need more.) If you use seaweed, you'll want to rinse away the salt first. If you're uncertain of the nutrient value of a mulching material, have your soil tested annually so you'll know what changes are taking place in your soil.

Inorganic Mulch

Inorganic mulches include those materials that don't improve soil, like plastic or gravel. Gravel isn't practical in most vegetable gardens, but black plastic is widely used since it warms the soil early for the fastest crops. Landscaping fabrics are similar to black plastic but have tiny pores to let water and air in. Gardeners might find

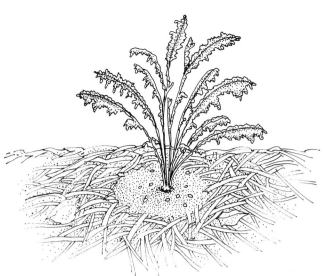

When mulching around plants, keep a 4- to 8-inch (10 to 20 cm) circle clear of mulch around your plants. This will minimize slug and snail damage, and disease problems.

both black plastic and landscaping fabric expensive, but they can last several seasons if you remove and store them each fall.

Black plastic should be put in place early in the season. Apply it several weeks before planting, to warm the soil. Landscape fabrics can go on anytime before planting. If you decide to use either of these materials, make sure you purchase the right width and length for your beds or rows. To determine the right width, measure across beds and rows then add at least 6 inches (15 cm) to each side. After you lay it out, anchor the sides with soil, rocks, or boards. Cut away holes for planting. At the end of the season, remove and store it until next season, if it's still in good condition.

Black plastic warms the soil and kills weeds by keeping them in the dark. Lay it across the bed, anchor the sides securely, and cut cross-slits to make planting holes.

Straw is a popular, long-lasting, and inexpensive mulch available directly from farmers or garden centers.

Pest Control and Mulch

Recent research suggests that mulches can help control pests in several ways:

- Mulch acts as a barrier between your plants and soilborne plant pathogens and insect pests.
- Mulch shelters beneficial organisms that help to control pests.
- A mulch of compost helps control soilborne plant disease, since beneficial decomposer organisms are stronger than many plant pathogens.
- A mulch of aluminum foil confuses aphids and keeps them from landing on and damaging plant leaves.
- A 6-inch (15 cm) layer of straw mulch has been found to help control Colorado potato beetles in potatoes.

Watering

Water makes up from 85 to 95 percent of the weight of living plants, so it's not surprising that an inadequate water supply causes them to stop growing and wilt. Plants use more water than they do anything else, and they need an adequate amount at the right time. Too much water is just as unhealthy for plants as too little.

When to Water

Your goal is to water *before* wilting occurs. Garden plants require about 1 inch (25 mm) of water each week. Gardens in hot, dry climates will lose moisture faster and may need up to 2 inches (50 mm) each week. In cool and wet climates, less water evaporates from the soil and you may not to have to water at all.

To monitor weekly rainfall, purchase a rain gauge (available at most hardware stores) and set it in or near your garden. Check it immediately after rainstorms, before any water evaporates. If weekly rains are inadequate, you should plan to water soon to maintain plant health and growth.

To check your soil for moisture, dig a small hole with a trowel and examine the soil in the root zone (the top 6 to 12 inches [15 to 30 cm] of soil). Sandy soils will flow freely through your fingers when dry, but will stick together slightly with adequate moisture. Heavier clay soils will appear hard and crumbly when dry, and

It's best to lay soaker hoses (dew hoses) with the holes facing down, between long rows of plants.

feel slick when adequately moist. If the soil is cool and moist, you can turn off the water and check the soil again a few days later.

How Much to Water

Water enough to keep your vegetables growing, but not so much that roots become oxygen-starved. For most soils, one good soaking is better than several shallow waterings. Use the guidelines given above to keep the soil at its optimum moisture level.

If you use an overhead sprinkler, monitor the rate of application by placing one or several rain gauges in your garden. Water for 20 to 40 minutes and check the soil again. Water until the root zone is soaked. The time required to water sufficiently will depend on your water

The advantage of small sprinklers is that they can be placed directly where water is needed.

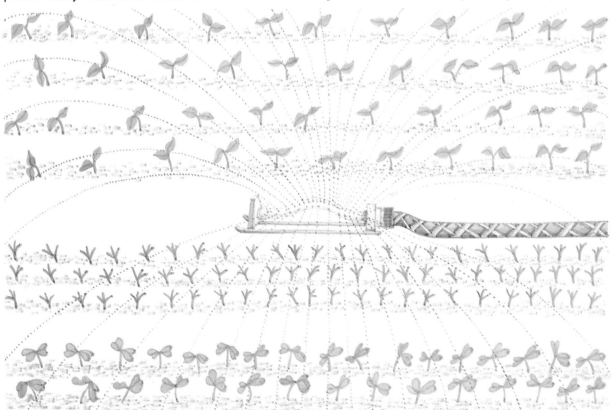

pressure, the size of the nozzles, the distance from the pump, the diameter of your hose, your soil type, and the drying effect of wind and sun.

How to Water

Rain barrels and sprinkling cans work well if your garden is small and you have plenty of time. If you're tending a large garden, overhead sprinklers and drip irrigation systems will save time. But if your goal is to conserve water as well as time, drip irrigation is best.

Overhead watering is easy and cheap. However, in the time it takes water to reach the soil, 30 to 50 percent may be lost to evaporation on a hot, windy day. Sprinklers take longer to wet the soil, especially if water must first penetrate a mulch. Fungal diseases that thrive on moisture spread easily and quickly when foliage is wet. When using a sprinkling system, water in the morning or early evening for 40 minutes two or three times each week in the absence of sufficient rainfall. Check the soil at root-depth to be sure you're watering enough.

Drip irrigation uses less water, since it is applied directly to the soil, where plants need it. You can water

Hand-watering is a relaxing task at the end of the day. It also gives you the opportunity to inspect your plants.

more efficiently in less time since almost no water runs off or evaporates. The water stays cool and helps keep the soil temperature low, especially if you mulch. And since foliage remains dry, fungal diseases are less threatening. The disadvantage is that drip systems, using water emitters, pressure regulators, and timers, are more expensive and require installation. Once they're installed, however, you'll only need to turn on the faucet to water your garden.

If you're considering purchasing a drip-irrigation system, start out small and add as you grow. Use a map of your garden to help during the planning stages. Most companies that sell drip-irrigation equipment have starter kits and will design a system to fit your needs and estimate the amount of water it will deliver.

You can design your own much simpler system by using soaker hoses. Also known as dew hoses, these are much less expensive than emitter drip systems but provide many of the same benefits. Some soaker hoses release water over their entire length, while others spurt water through tiny holes. Make sure the holes face down into the soil when using this type of hose. A soaker-hose system needs no assembly—simply lay the hoses between plants and along rows where you need them. The hoses can be made of various kinds of plastic, rubber, or canvas. Some rigid hoses can be difficult to manipulate exactly where you want them, but they seldom develop leaks and, unlike canvas hoses, they're not susceptible to mold and mildew.

How to Conserve Water

Water is a natural resource that must be conserved, preserved, and protected, so use only what you need. Here are a few tips:

- Designate sections of the garden for vegetables with low, medium, and high water requirements and water them individually. Refer to the "Plant by Plant Guide," starting on page 78, for the crops you wish to grow.
- Use a rain barrel to catch water running from your roof. You can use a length of hose to siphon the water to your garden.
- Insulate the soil surface with a thick layer of organic mulch. To learn how to do this, read "Mulching" on page 64.
- Maintain your soil's organic matter by working in plenty of compost or other organic materials each season. Organic matter holds water like a sponge.
- Control weeds, since they compete with your vegetables for moisture.
- In arid climates, it is very important to maintain your soil's water content by mulching often, adding organic matter, and providing wind protection for plants.

Weeding

Weeds are often the most visible and persistent pests. Weeds are simply plants growing in the wrong place at the wrong time. Most of them are aggressive wild plants that know a good opportunity when they find it—and if you're unprepared, they'll find it in your garden.

The Role of Weeds

Many of the plants we call weeds have culinary or medicinal uses that were discovered long ago by indigenous populations. The tender leaves of dandelion and lamb's-quarters, for example, can be used in salads along with the garden lettuce. Wild comfrey was used medicinally long before modern herb growers added it to their gardens.

Weeds hold the soil in place, break up compacted soil with their vigorous root systems, and conserve nutrients that would be leached away if the soil were left bare. Many beneficial insects that help to control garden pests spend part of their life on weeds. Plants like Queen Anne's lace and goldenrod, which are generally thought of as weeds, are dependable bloomers that provide nectar for beneficial insects.

Some farmers use weeds as indicators of soil imbalances. Dandelions and quack grass, for example, prefer soil that is compacted and low in oxygen—think of them as red flags for areas that need extra organic matter. Mustards and sorrel flourish in acid soil, meaning it's time to have the soil pH tested.

Weeds are pests if they compete with your vegetables for nutrients, moisture, and light. They harbor diseases or pests that can move to your vegetables. Some weeds, like horse-nettle and nightshade, are poisonous. Others, like poison ivy, can cause allergic reactions.

How to Control Weeds

You should first identify the weeds you need to control. Learn to recognize annuals and perennials so that you'll know whether to till them in (the annuals) or pull them out (the perennials). Learn to distinguish between vegetable and weed seedlings, so you don't destroy your vegetable seedlings by mistake.

Bindweed is an attractive plant with an unattractive habit of entwining itself around your other garden plants and choking them.

The annual weeds live only one season but long enough to produce millions of seeds that can remain dormant in your garden soil as long as 100 years, coming to life when you till a new garden patch. Pulling, hoeing, and cultivating are often sufficient to control them. The summer annuals, like crabgrass, giant foxtail, pigweed, and lamb's-quarters, begin life in spring and die in fall. Winter annuals, like chickweed and yellow rocket, sprout in late summer and survive the winter in dormancy, flowering and reseeding themselves the following spring before they die. Biennials, like Queen Anne's lace, live for two seasons, flowering and dying during their second year. Control all annual and biennial weeds before they flower.

Perennial weeds live more than 2 years, and are pesky because they can reproduce sexually by seed, or vegetatively with special root and stem structures. Don't till perennials like quack grass and field

Canada thistle, with its pretty purple flowers and prickly leaves, can become a serious weed pest.

A hand fork is useful for pulling out stubborn weeds, particularly in a patch of rough, dry ground. It is the tool to use when sitting or kneeling on the ground.

The stubborn deep roots of bindweed need close attention with a hand fork in order to really get them out.

Lamb's-quarters, also known as fat hen and white goosefoot, is a fast-growing weed with edible leaves that can be eaten in salads. You can also cook it the same way you do spinach.

bindweed into the soil, since this will chop and distribute their hardy root pieces, which sprout new plants.

Eliminate Weed Seedlings Scrape the soil surface with a hoe each week to eliminate weed seedlings. Choose any blade, but make sure that the handle is long enough to allow you to stand straight; a short-handled tool may cause back strain. Your motion should be more like sweeping with a broom, rather than like chopping.

Prevent Seed Formation Pull older weeds before they have a chance to flower and make seed.

Create a Weed Barrier Mulch with synthetic or organic materials. Apply a mulch of straw or compost at least 4 to 12 inches (10 to 30 cm) deep, as it will settle to a shallow layer over time. In dry climates, mulch early to retain soil moisture. In wet climates, allow soil to dry before covering it for the season. See "Mulching" on page 64 for complete directions on how to mulch.

Till Shallowly When cultivating between rows, or after solarizing, till only the top 1 inch (2.5 cm) of soil, to avoid bringing up dormant weed seeds that will germinate when exposed to light.

Grow Cover Crops Sow cover crops at any time of year. Crops like buckwheat grow best when it's warm. Rye and legumes like it a bit cooler. A good cover crop grows quickly and out-competes native weed populations. To learn how to use cover crops, see "Green Manures" on page 44.

Solarizing Your Soil

Before planting, you may want to solarize the top layer of your garden soil. Midsummer is the best time to solarize soil because the method works well only if the weather's consistently dry and hot. Cultivate and remove all crop debris from the soil, rake it, and water it if it's dry. Then lay a sheet of clear plastic over the soil and press it firmly against the soil. Tuck the edges in a shallow trench around the bed, then cover them with soil to seal. Left for several weeks, the soil under the plastic will heat and kill most weed seeds at the surface.

Often thought of as a weed, clover is actually a legume that can add nitrogen to the soil. Clover is also often used as a cover crop.

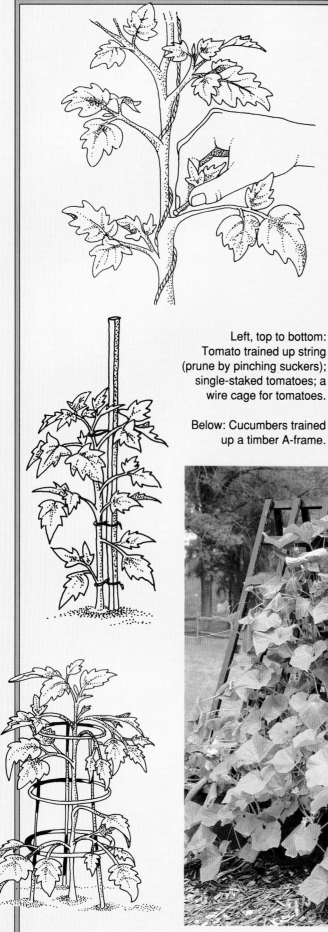

Training Methods

Training vegetable plants to grow vertically, instead of horizontally, saves space and makes cultivating and harvesting easier. Staking improves plant health, since better air circulation promotes drying and lessens the chance of disease. Since fruits aren't in contact with soil, they're less likely to succumb to soilborne pathogens. Staked plants are more attractive, and leave you more room for special techniques like interplanting and companion planting.

You can purchase stakes made of wood, metal, plastic, or bamboo in a variety of heights to fit the needs of the vegetable. Sturdiness is more important than appearance, since vegetable plants take up lots of space and become heavy with fruit as the season progresses.

No matter what method you use, have your support system in place before you plant. Vines like peas and beans will attach themselves by wrapping tendrils or twining stems around the support. Plants like tomatoes or peppers need some help: Tie them loosely to the support, using scraps of soft cloth or cotton rope.

Left, top to bottom:
Tomato trained up string
(prune by pinching suckers);
single-staked tomatoes; a
wire cage for tomatoes.

Below: Cucumbers trained
up a timber A-frame.

Universal Plant Cages

You can make long-lasting, universal cages from concrete reinforcing wire for supporting tall plants like

tomatoes and peppers, as well as vines like cucumbers, pole beans, and peas. Buy the wire at building-supply stores in rolls 100 feet (30 m) long and 6 feet (2 m) high. The mesh is large—each square measures 5¼ x 6 inches (13 x 15 cm), big enough for reaching through to harvest most vegetables.

You'll need bolt cutters and a pair of pliers. Cut off an 8-foot (2.4 m) section for each cage, making your cut through the middle of the sixteenth row of squares. Wrap the wire around to form a cylinder, using the cut ends of wire as fasteners. To use the cage for tomatoes or peppers, stand the cylinder upright. You can cut away the horizontal wire on the bottom, leaving 5-inch (13 cm) prongs to insert into the soil as anchors. For added strength in windy climates, drive a 4-foot (1.2 m) stake into the soil 12 inches (30 cm) deep next to the cage and another on the opposite side. Use twine or wire to fasten the cage to the stakes. Place one plant in the center of each cage. As the plants grow, they'll fill the cage with foliage as they push themselves upward.

You can also grow vining crops like cucumbers or gourds with the cage upright, or turn the cage on its side and secure it with several stakes. Plant inside the cage; the vines will loop around and form a horizontal tube. This works well for heavier crops like melons.

Right, top to bottom: Simple A-frame supports for tomato plants; vertical trellis for cucumbers; beans trained up a teepee.

Below: Timber framing is used for raised beds as well as vertical trellising.

Controlling Pests and Diseases

The best pest controls are specific, safe, easy, and effective. If your garden plants are bothered by pests, choose control methods that target the pests and not other organisms. Select controls that are safe for you and the rest of the environment. They should be easy to use and provide the control you need.

Proper identification is the first step toward successful pest control. Consult a book on identifying garden pests, or submit insect or disease samples to your local Cooperative Extension Service. Identification is important because not all pests respond the same way to control practices.

Resistance

One of the easiest ways to avoid problems is to choose cultivars genetically designed to resist the pests that plague your vegetables. Plant breeders have developed an array of new cultivars able to defend themselves against specific diseases and insect pests. Find out which conditions will be a problem and look for resistant cultivars in seed catalogs and garden-supply stores.

Plant Environment

Healthy, vigorous plants can withstand light pest attacks, so create the best garden environment above ground and below ground. Test your soil every 3 to 5 years, and use the results to guide you in balancing plant nutrients in the soil. Make yearly additions of compost to maintain soil organic matter and to suppress plant pathogens. Monitor weekly rainfall and soil moisture to make sure your plants receive the right amount of water, and begin irrigating when necessary.

Weekly Inspection

Check plant leaves, flowers, and stems for signs of insects or pathogens at least once each week. In large stands, examine three to five successive plants at three different locations. Become familiar with insect life cycles so you'll recognize their resting (egg and cocoon) and active (larva and adult) stages. Most insects you find are likely to be beneficial or harmless. If you find pest insects, pick them off by hand early in the cool morning, when they're sluggish. You can use a portable vacuum to suck them away, or knock them from plants with a forceful spray of water.

Sanitation

When buying transplants, reject those with symptoms of disease or insect damage. Healthy specimens should be vibrant, green, and bushy. For more information on choosing healthy seedlings, see "Buying Plants and Seeds" on page 34.

If isolated pest problems develop, sacrifice the plant rather than risk a garden-wide infection. Pull and dispose of infested plants by burying them deeply, burning them, or placing them in a fresh, hot compost pile. To get rid of insect eggs, cocoons, and adults, smash them, drown them in kerosene or soapy water, or freeze them in bags before disposal.

While insect problems can be limited to a few plants, diseases often spread quickly to affect your whole crop. Following a few simple routines will help to contain an outbreak. When working among disease-infested plants, rinse hands, tools, and boots in a 10 percent bleach solution (1 part bleach to 9 parts water) immediately afterward to avoid spreading disease organisms to healthy plants. Several times each year, collect garden refuse and use it to build a hot compost pile. (For instructions on how to make a hot compost pile, see "How to Make Fast, Hot Compost" on page 46.) This will help to destroy the overwintering sites needed by the disease organisms, and to keep them from lingering to attack next year's crop.

Barriers

Keep insect and other pests out with impenetrable barriers. Spun-bonded floating row covers are widely

Aphids

Colorado potato beetle larvae

Cabbageworm damage

Not all insects are bad. The appetite of the insect-eating praying mantis makes it a natural pest controller, and therefore a welcome guest in the vegetable garden.

available. Use them to cover rows, beds, or the whole garden, anchoring the sides with soil or stakes. Put them in place immediately after planting. Leave enough slack for plants to push the covers up as they grow. Remove them when it's time to weed, or when temperatures get too hot for cool-season crops. If you're growing crops that are insect-pollinated (like squash), remove the covers when the plants flower. Floating row covers can be stored and reused for several seasons.

Traps

Pheromone traps lure insects away from your vegetables and break their mating cycles. Pheromones are chemicals, in the form of an odor, naturally emitted by female insects to attract a mate. Pheromone lures mimic this action, attracting males from miles around. You can purchase synthetic pheromones to use with sticky cardboard traps. Hang them in the garden. When males approach, they're trapped in the sticky coating and are unable to mate. Unmated females won't lay eggs and the next generation of pests will be depleted.

You can also use pheromone traps to monitor pests, so you'll know whether pest numbers are increasing, and whether more drastic control measures are necessary. Purchase pheromone lures and sticky traps from garden-supply stores and catalogs for pests like cabbage looper, corn earworm, European corn borer, fall armyworm, and others. Follow the manufacturer's instructions for assembly and placement.

Biological Controls

In natural environments, pests are just a small part of the complex diversity of animals and organisms that live on or near plants and soil. Actually, pests are outnumbered by insects and microorganisms that help keep your garden productive (like bees that pollinate, and microorganisms that turn organic wastes into compost). The community of organisms that makes your garden home includes "beneficials" that prey upon the pests, like parasitic wasps, predaceous insect-eating bugs, toads, and frogs. The use of living organisms, like beneficials, to control pests is called biological control. Other forms of biological control include microbial insecticides and beneficial nematodes (microscopic worms). Microbial insecticides carry specific insect diseases which cause insects to get sick; BT (*Bacillus thuringiensis*) is an example of a microbial insecticide.

Pests and beneficials regulate their own numbers and keep each other in check, but gardening can upset the natural balance. That's when you notice damaged leaves or fruit, or symptoms of disease like soft spots or withered leaves. It's a good idea to encourage the beneficials by creating the right environment. Use plenty of compost for everything, everywhere in your garden. Mix flowering plants and herbs among your vegetables, to provide food and shelter for beneficials. Avoid using botanical pesticides, since they take their toll on beneficials as well as pests. For more information on attracting beneficial insects, read "Companion Planting" on page 52.

Slug damage on radish

Slug eggs

Red spider mite damage on cucumber

You can supplement the beneficials in your garden by purchasing specific biological controls from garden-supply catalogs and insectaries. Releasing biological controls works well if you've identified the pest correctly and purchased the right control. If you use them, follow the instructions that arrive with the package for storage and release instructions.

Plant Disease

Disease symptoms are caused by pathogens, including fungi, bacteria, and viruses. Prevention is your best defense against plant pathogens, since once they infect plants they're almost impossible to eradicate. Proper sanitation is the best prevention. If you've had problems in past years, you can try to prevent infection by most fungi and some bacteria by treating plants every 5 to 10 days with an organically acceptable fungicide like copper or sulfur. Purchase these products at garden-supply centers and follow the directions on the label when using. Try also the homemade remedies suggested in "Homemade Pest Controls" below.

Homemade Pest Controls

Experiment with homemade pest controls. First, make test applications by treating just a few leaves and waiting a few days; if plants are damaged, dilute the mix with water, or refrain from using it on sensitive plants. Keep equipment for making pest controls separate from kitchen utensils. Most home remedies aren't sure-fire cures—you'll have to monitor plants to check the effectiveness.

Baking Soda Spray plants with a solution of 1 teaspoon baking soda mixed with a few drops of liquid dish soap and 1 quart (1 l) water, to control fungal disease. It prevents spores from germinating, and may also help stop infections already started.

Compost Tea Mix one part compost with five parts water; allow to stand for 1 week, then strain.

You can mix up some very effective pest controls right in your kitchen. A mixture of garlic, oil, soap, and water is a natural control for insects and plant disease.

Spray this "tea" onto plants to prevent fungal disease.

Garlic Soak 3 ounces (85 g) minced garlic in 2 teaspoons of mineral oil for 24 hours; strain. Add 1 pint (600 ml) water and 1 teaspoon liquid dish soap; mix thoroughly. Spray plants with a solution of 1 to 2 tablespoons soap mixture and 1 pint (600 ml) water, to control insects and disease.

Aromatic Remedies Add several drops essential oil such as rosemary, pine, or citronella to 1 cup (8 fl oz/ 250 ml) water; spray plants to control insect pests.

Bleach Disinfect tools and seed-starting supplies with a 10 percent solution of household bleach (sodium calcium hypochlorite).

Soap and Oil Add 1 tablespoon liquid dish soap to 1 cup (8 fl oz/250 ml) oil (peanut, safflower, corn, soybean, or sunflower). Mix 1 to 2½ teaspoons of the soap-and-oil base to 1 cup (8 fl oz/250 ml) water; spray plants to control a wide range of insect pests.

Powdery mildew on pea leaves

Rust on fava beans

Scab on potatoes

Common Vegetable Pests and Diseases

Pest	Damage	Prevention and Control
Snails and Slugs	Seedlings eaten, irregular holes in leaves.	Place shallow containers of beer in garden, or trap pests under boards.
Aphids	Foliage wilted or curled, deformed buds and flowers.	A short, sharp spray of water from the hose will dislodge them. Spray with insecticidal soap.
Cabbageworms	Leaves eaten.	Use row covers. Spray or dust with BT.
Carrot Fly	Carrot roots eaten.	Rotate carrot plantings. Use row covers. Plant crops in late spring to minimize damage.
Cutworms	Plant stem chewed at soil surface.	Place cardboard or metal cutworm collars around the stem of the plants. Sprinkle moist bran mixed with BTK on the soil surface in the evening. Add parasitic nematodes to the soil at least a week before planting.
Leaf Miners	Winding or large blotchy lines on leaves, especially beets, spinach, and tomatoes.	Use row covers. Pick and destroy infected leaves. Control adults with yellow sticky traps or a pyrethrin/rotenone mix.
Squash Borer	Holes in base of stem, wilting leaves.	Use row covers until flowers appear.
European Corn Borer	Tunnels in corn stalks and ears.	Use pheromone lures and sticky traps.
Beetles (flea, Mexican bean, Japanese, cucumber)	Chewed foliage.	Eliminate weeds. Use row covers. Spray with garlic or pyrethrin.
Red Spider Mite	Discoloration and bronzing of the foliage.	Use a garlic or pyrethrin/rotenone spray.

Disease	Damage	Prevention and Control
Powdery Mildew	Downy patches on foliage.	Provide good air circulation. Control weeds. Spray foliage with compost tea.
Damping-off	Seedlings weaken and collapse because of decay at soil line.	Start seeds in well-drained mix. Avoid over-watering, crowding, and poor air circulation. Disinfect reused pots and flats.
Rust	Rust-colored powder on leaves.	Provide good air circulation. Remove infected leaves.
Wilt (Fusarium and Verticillium)	Leaves yellow, plant gradually wilts.	Rotate crops. Plant resistant cultivars. Destroy infected plants.

Chicory roots are harvested in fall and may be stored for up to 18 weeks in damp soil in the cellar.

Harvesting and Storing

How you pick and store your harvest influences vegetable quality and flavor. Check the "Plant by Plant Guide," starting on page 78, for tips on harvesting and storing specific vegetables.

When to Pick

Harvest vegetables early, often, and at their peak tenderness and flavor. Quality declines with time, so you'll sacrifice flavor, texture, and storage quality the longer you neglect to pick. Frequent harvests prolong the productive lifespan of most vegetable plants—the more you pick, the more the plant produces, since harvesting delays or prevents the production of seed. After producing seed, many vegetables stop producing or die.

Mark the expected date of maturity for your crops on your garden calendar, and begin inspecting vegetables for ripeness at least 1 week before this date. In your garden, vegetables may mature earlier or later than expected due to the influence of climate or pests. For example, unseasonably warm weather can hasten a harvest; or pests may weaken plants and delay ripening.

Pick vegetables in the early morning while they're still cool, but after the dew has dried. Flavor and nutrition are at their peak when you pick, so aim for using vegetables soon after harvest.

If you're growing "baby" vegetables, plan on picking every day for the best quality and production. You can pick baby leeks and onions when they're the size of a pencil. Baby squash should have even color—you'll know that you've picked too early if the flavor is bitter. Select baby carrots that have good color and flavor. Many carrot cultivars grown for their large roots aren't suitable as tiny roots since they don't develop that deep,

Storing Fresh Vegetables

Refrigerate and use within several days to 1 week: artichokes, asparagus, beans, broccoli, collards, eggplant, endive, kale, lettuce, mushrooms, mustard greens, okra, peas, radishes, spinach, sweet corn, Swiss chard.

Refrigerate and store up to 12 weeks: beets, brussels sprouts, cauliflower, celery, leeks, melons, onions (green), peppers, radicchio, summer squash.

Keep cool and damp up to 18 weeks (32 to 40°F/0 to 4°C and 90 percent relative humidity): cabbage, carrots, kohlrabi, parsnips, potatoes, rutabagas, turnips.

Keep cool and dry up to 7 months (32 to 40°F/0 to 4°C and 65 to 70 percent relative humidity): garlic, onions (except green).

Keep warm and dry up to 2 to 3 months (50 to 55°F/10 to 12°C and 50 to 70 percent relative humidity): pumpkins, winter squash.

orange color until later. You can choose vegetable cultivars bred especially for immature production; they'll offer the best quality at an early age.

Handling

Brush away the soil clinging to roots or leaves, but don't wash vegetables before storing: Adding moisture just seems to invite spoilage. Pull away bruised or damaged parts before you store and add them to the compost pile.

Most vegetables grown for long-term storage should be "cured" before they're tucked away. The best way to cure garlic and onions is by hanging them in braids or mesh bags, in a dry and airy place out of direct sun—barn or attic rafters are ideal. Leave them to dry for several weeks. You can cut away their tops, but leave several inches of stem.

When you cut pumpkins and winter squash from the vine, leave several inches of stem. Leave them in the field for 2 weeks after picking, to toughen their skins. If freezing weather arrives early, bring them indoors and keep them at about 70°F (21°C) for the same period. Potatoes will store best if they're held at 50 to 60°F (10 to 15°C) and high relative humidity for 10 to 14 days before storing.

Storing

When it's time to store, sort your harvest into two groups: vegetables that require moisture, and those that require dryness. Use "Storing Fresh Vegetables" on page 76 to help you choose the best storage method for each crop.

In cool climates, you can leave crops like carrots, leeks, and turnips in the ground over winter (in warm climates, vegetables left in the ground will attract insect pests). Cover with at least 6 inches (15 cm) of mulch; in severe climates, cover the row with bales of straw. Push aside the covering to harvest the roots. Or place harvested vegetables in a single layer sandwiched between 6-inch (15 cm) layers of mulch. Build several layers, then top with 2 feet (60 cm) of mulch and cover with a plastic tarp. Other options for winter storage include empty growing (cold) frames and root cellars. The ideal spot is cold and dark but not freezing.

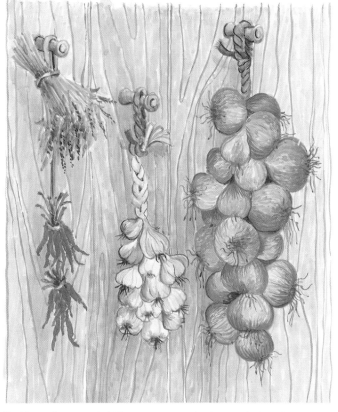

Garlic, onions, and chili peppers may be harvested and hung in a dry, airy place out of direct sunlight.

Potatoes should be harvested before a hard freeze. Dry without washing and store in well-ventilated boxes or mesh bags in a cool to cold area.

Pumpkins and winter squash may be stored for up to 2 to 3 months in a warm (50 to 55°F/10 to 12°C), dry cellar.

Preservation

If simple storage methods don't fit your needs, try canning, freezing, or drying your vegetables. No matter which method you choose, use only fresh, top-quality produce. Most suppliers of food-preservation equipment include instructions with their products. You can also consult your local Cooperative Extension Service for the latest techniques on safe food preservation at home.

Be sure to use the right method, since vegetables and even cultivars respond to preservation differently. The best vegetables for canning include asparagus, beets, beans (green and wax), carrots, corn, hot peppers, okra, onions, peas, potatoes, pumpkins, rhubarb, tomatoes, and winter squash. Good choices for drying include beans, beets, carrots, celery, garlic, mushrooms, onions, parsnips, peppers, pumpkin, tomatoes, and summer squash. Vegetables recommended for freezing include asparagus, broccoli, brussels sprouts, carrots, cauliflower, corn, lima beans, peas, peppers, summer and winter squash, and tomatoes made into sauce.

PLANT BY PLANT GUIDE

All you need to know about planting, maintaining, and harvesting a wide variety of vegetables is found in this "Plant by Plant Guide." And to make matters easier, the guide is arranged in an easy-to-use, quick-reference format. (See "How to Use This Book" on page 10 for details on how to use the "Plant by Plant Guide.")

The Plant by Plant Guide will ensure that you have success with your vegetable garden. All the vegetables are listed in alphabetical order by common name. Each vegetable is also illustrated with a full-color photograph, and each entry supplies specific propagation and cultivation details.

The entries include information on the climatic zones—where the plant grows and whether it prefers sun or shade. Refer to the USDA Plant Hardiness Zone Map on page 154 to find out which zone you live in, since it is important to choose plants that grow in your climatic zone.

There is also information on ideal soil conditions for each plant, including the preferred pH range (a measurement of the soil's acidity or alkalinity). Also see "Soil" on page 22 to find out how to adjust the pH of your soil.

The "Growing Guidelines" tell you when to sow seed, at what stage to transplant the vegetables into the garden, when to feed and mulch, and how much to water your vegetables.

The entries also tell you how to deal with pests and diseases using organic methods. They also provide information about common problems. Having knowledge of the common problems of each vegetable is helpful in preventing pests and diseases.

In addition, there is information on how many days the vegetable takes to reach maturity and how to harvest and store each vegetable. "Special Tips" covers many topics, such as companion planting and how to get a second harvest from your vegetables. There are many other details that make this a complete guide for happy and successful vegetable growing.

Opposite: Golden beets are prized for their unusual bright orange color. Beets thrive in almost any climate. They should be grown in well-drained, rich, neutral soil, and can be harvested as soon as large enough for use.

| *Amaranthus tricolor* | Amaranthaceae | *Cynara scolymus* | Compositae |

AMARANTH

ARTICHOKE, GLOBE

Grow amaranth for its nutritious leaves, which can be eaten both raw in salads and cooked as greens.

This plant is grown chiefly for its large, edible flower heads, but suckers also may be blanched and eaten like asparagus.

BEST CLIMATE AND SITE: Zones 4 and warmer, with protection in colder areas. Prefers heat and full sun.

IDEAL SOIL CONDITIONS: Not fussy, but prefers rich, humusy soil; pH 6.0–7.0.

GROWING GUIDELINES: Sow seed after frost, when soil has warmed. Barely cover and keep moist until germinated. In short-season areas, start amaranth indoors 4 weeks before last frost date. Water freely and thin to 18 inches (45 cm) apart.

PEST AND DISEASE PREVENTION: Protect plants from flea and cucumber beetles with spun-bonded row covers, or spray with garlic infusion. Rotate plantings to avoid stem rot from soilborne diseases.

COMMON PROBLEMS: In cooler climates, amaranth may require soil-warming plastic mulch.

DAYS TO MATURITY: Requires 40–55 frost-free days, although thinnings may be harvested earlier.

HARVESTING AND STORING: Harvest whole plants as thinnings; pick the young, tender leaves from mature plants. Harvest frequently to encourage new growth. May be frozen like spinach.

SPECIAL TIPS: Plant away from cabbage-family plants or cucumbers to reduce pest damage; or plant adjacent to those crops to serve as a "trap crop."

CULTIVARS: No named cultivars, but oriental seed companies often carry natural variants with light green, dark green, or red-striped leaves.

OTHER COMMON NAMES: Chinese spinach, Joseph's-coat, leaf amaranth, tampala, vegetable amaranth.

BEST CLIMATE AND SITE: Mild coastal climates, Zones 8–9; can be grown in Zones 5–7 with special winter care. Full sun and protected exposure.

IDEAL SOIL CONDITIONS: Rich, well-drained soil; near-neutral pH.

GROWING GUIDELINES: Start seed indoors in late winter and transplant in late spring, when soil has warmed thoroughly. Can reach 5 feet (1.5 m) or more in height; space seedlings 24 inches (60 cm) apart in the row and allow 3 feet (90 cm) or more between rows. Be generous with compost or manure. Water deeply; mulch between rows.

PEST AND DISEASE PREVENTION: Little troubled by insect pests but is susceptible to crown rot. Well-drained soil is essential. Do not allow mulch to smother the crown.

COMMON PROBLEMS: In Zones 7 and colder, winter-kill is the chief problem. Cut the plant back in late fall and protect over winter by inverting a basket or box over the crown and mulching deeply.

DAYS TO MATURITY: May produce edible buds the first year but more likely will not yield a harvest until the second year. Buds appear in late spring in warm climates, in summer in cooler ones.

HARVESTING: Cut buds before the scales have begun to open, with 1 inch (2.5 cm) of stem attached.

SPECIAL TIPS: Side-dress with bloodmeal or other high-nitrogen fertilizer in early spring.

CULTIVARS: 'Green Globe' is widely available.

Eruca vesicaria subsp. _sativa_   Cruciferae

ARUGULA

The tender leaves of this European garden favorite add a nutty, peppery bite to salads and sandwiches.

BEST CLIMATE AND SITE: All Zones. Full sun as well as half-day sun or partial shade. Avoid hot, dry positions. In mild climates, grow as a winter vegetable.

IDEAL SOIL CONDITIONS: Not fussy, but prefers fertile, moist soil; pH 6.0–7.0.

GROWING GUIDELINES: Direct-seed as early as possible in spring. Light frost will not harm the seedlings. Thin to 8 inches (20 cm) apart, using thinnings in salads.

PEST AND DISEASE PREVENTION: Cover with spun-bonded row cover to deter flea beetles.

COMMON PROBLEMS: Sown too late, spring-planted arugula will bolt (go prematurely to seed) in warm weather before reaching harvestable size. In cold climates, sow seed indoors and set out as soon as ground can be worked. A fall crop, direct-seeded or set out as seedlings 1–2 months before the first fall frost, will stand longer without bolting. Extend the harvest with successive plantings a week apart.

DAYS TO MATURITY: 40 days, although thinnings may be harvested earlier.

HARVESTING AND STORING: Pick large leaves from bottom of plant. New leaves will sprout from center crown. Weekly picking should yield 8–12 large leaves per plant. Use fresh; do not freeze.

SPECIAL TIPS: Add flowers of bolting plants to salads.

OTHER COMMON NAMES: Rocket, roquette, rucola, rugula.

Asparagus officinalis   Liliaceae

ASPARAGUS

This classic spring vegetable requires well-prepared soil with high fertility. A well-maintained patch may yield for decades.

BEST CLIMATE AND SITE: Zones 3 and warmer, but asparagus grows best in areas where soil freezes in winter. Avoid low-lying areas subject to heavy dew and morning fogs, to reduce the possibility of rust. Prefers full sun, but will tolerate some shade.

IDEAL SOIL CONDITIONS: Fertile, well-drained soil; will tolerate slightly alkaline pH (6.5–6.8) and saline soils.

GROWING GUIDELINES: Grow from seed started indoors or in an outdoor seedbed; or hasten the first harvest by using year-old crowns. Dig a trench 8 inches (20 cm) deep in well-composted and well-limed soil. Place the crowns in the trench 15 inches (38 cm) apart, fanning the roots in all directions. Cover with soil to half the depth of the trench. When foliage peeks above ground level, finish filling the trench with soil. Mulch or cultivate shallowly and irrigate in dry spells. Keeping the foliage healthy and lush after harvest is critical to the next year's crop. Each fall, cut back dead foliage and mulch heavily with compost or strawy manure. Early each spring, rake off all but 1–2 inches (2.5–5 cm) of mulch to let spears emerge.

PEST AND DISEASE PREVENTION: Asparagus rust can be a serious problem in damp locations; use rust-resistant cultivars. Reduce damage from asparagus beetles, which overwinter in garden debris and emerge in spring to feed on young spears, by burning or hot-composting the old asparagus

ASPARAGUS—CONTINUED

ASPARAGUS PEA

When the harvest is finished the graceful fronds of asparagus add an ornamental touch to the garden.

Unique flavor and high protein content make the asparagus pea popular in Asian cuisines. It also produces edible tubers.

foliage and cultivating the asparagus patch shallowly before applying fall mulch.

COMMON PROBLEMS: Perennial weeds and grasses can be troublesome in the asparagus patch. Be sure the area is free of perennial weeds before planting, and mulch or cultivate to stay ahead of them.

DAYS TO MATURITY: Gardeners were once advised to wait until the third year to take their first small asparagus harvest; newer hybrid cultivars may be picked sparingly for 1–2 weeks in the second year. A mature patch, perhaps 5 years old, may be harvested for as long as 10 weeks, yielding 1 pound (500 g) or more of asparagus for each foot (30 cm) of row.

HARVESTING AND STORING: Carefully cut spears at ground level or simply snap them off, leaving any woody stem behind. Harvest while tips of spears are still tightly closed; in warm spring spells, this may require daily harvesting. Can or freeze excess spears.

SPECIAL TIPS: In heavy soils, plant in raised beds to improve drainage. Lay crowns at ground level, rather than in trenches, and mound soil over them.

CULTIVARS: 'Jersey Centennial', a high-yielding hybrid; 'Martha Washington' and 'Mary Washington', rust-resistant old favorites; 'Viking KB3', recommended for extremely hot or cold climates.

BEST CLIMATE AND SITE: Zones 7 and warmer; perennial in tropical climates. Full sun near a trellis or fence.

IDEAL SOIL CONDITIONS: Loose and well-drained soil; pH 7.3–8.0.

GROWING GUIDELINES: Plant after frost danger is past, 1 inch (2.5 cm) deep and 2–4 inches (5–10 cm) apart. The vines grow 6–8 feet (1.8–2.4 m), so provide a sturdy trellis or fence. Cultivate shallowly or mulch. The purplish red flowers form in loose clusters. The pods are four-sided, rather than round, with distinct flanges or "wings" at each corner.

PEST AND DISEASE PREVENTION: Do not grow where legumes, such as peas and beans, have grown the previous year. Keep vines growing strongly with regular watering to deter aphids.

COMMON PROBLEMS: Requires a long, warm growing season. Where the weather is not to its liking, it may die without flowering.

DAYS TO MATURITY: 120–150 frost-free days for edible pods; 180–210 days for tubers.

HARVESTING AND STORING: Harvest the pods at 6–8 inches (15–20 cm) and eat steamed or sautéed. Pick frequently, as pods can quickly become oversized. Dig marble-sized tubers after frost has killed the plants; use in stir fries or stews.

CULTIVARS: Named cultivars not generally available.

OTHER COMMON NAMES: Goa bean, Manila bean, princess pea, winged bean.

Phyllostachys spp.　　　　　Gramineae

BAMBOO SHOOT

Where climate and garden space allow, bamboo provides edible young shoots each spring as well as useful garden stakes later.

Delicious young bamboo shoots may be added to Asian dishes or used fresh in salads.

BEST CLIMATE AND SITE: Zones 6–9; hardiness varies among species. Full sun.

IDEAL SOIL CONDITIONS: Fertile, moist but well-drained soil; pH 6.0–7.0.

GROWING GUIDELINES: Propagate by division in late winter or very early spring. Buy potted divisions by mail or in garden centers in areas where bamboo is hardy. Plant in early spring and keep watered until well established.

PEST AND DISEASE PREVENTION: Little troubled by pests.

COMMON PROBLEMS: Most bamboos that produce edible shoots are "running" bamboos that spread by underground rhizomes. Build a strong underground barrier or mow frequently to keep the colony from getting out of hand.

DAYS TO MATURITY: Depending on vigor of the species, some shoots may be harvested the second spring.

HARVESTING AND STORING: Cut tender young shoots at ground level in early to midspring. Stir fry or use fresh in salads. Cut stalks for use as garden stakes when they are at least 3 years old.

RELATED PLANTS: Edible species of *Phyllostachys* vary widely in their hardiness and mature height. The smallest is *P. aurea,* also called fishpole or golden bamboo, which grows to about 20 feet (6 m) tall at maturity. At the other extreme is *P. bambusoides,* or timber bamboo, which can grow to 72 feet (22 m) tall and 6 inches (15 cm) in diameter. Some of the hardiest bamboos include *P. aureosulcata, P. dulcis, P. elegans,* and *P. nuda,* which grow to 30–35 feet (9–10.5 m) tall. *Phyllostachys* species do not do well in tropical or subtropical areas, but in Zone 10 you can grow *Bambusa beecheyana,* an edible bamboo that grows to 50 feet (15 m) tall and 5 inches (12.5 cm) in diameter. *B. beecheyana* is a "clumping" bamboo, which is easier to control than the running types.

Phaseolus vulgaris   Leguminosae

BEAN (DRIED)

The hyacinth bean produces delicious pods and seeds. You can also use the purplish flowers in salads and dips.

A lover of warm climates, the black-eyed bean is simple to grow and rich in protein. Add it to vegetable soup.

BEST CLIMATE AND SITE: Zones 4 and warmer; use quick-maturing cultivars in colder areas. Full sun, in a site with good air circulation.

IDEAL SOIL CONDITIONS: Well-drained garden soil is acceptable; pH 6.0–7.0.

GROWING GUIDELINES: Sow seed after last frost and when soil has warmed, 1 inch (2.5 cm) deep and 3 inches (7.5 cm) apart in single or double rows. Cultivate shallowly until the plants are large enough to shade out weeds. Mulch between rows to help prevent pods from rotting if they touch the ground. Moisture is critical when plants are in flower; when pods begin to mature, withhold water to hasten drying.

PEST AND DISEASE PREVENTION: Provide good air circulation to help prevent blights, mosaic disease, and anthracnose; to avoid spreading rust, do not disturb plants when foliage is wet. Till or spade under all plant debris in fall to destroy any disease organisms, and do not plant beans or other legumes in the same place more than once every 3 years.

COMMON PROBLEMS: Damp weather in late season, when pods are maturing, can encourage beans to sprout in the pod. Pull plants when most of the foliage has died and hang by the roots in a well-ventilated place to complete drying.

DAYS TO MATURITY: 90–150 frost-free days. The most popular garden cultivars are quick-maturing beans that are ready to harvest in 100 days or less.

HARVESTING AND STORING: Harvest when pods are completely dry and beans can barely be dented when bitten. Shell pods individually or thresh by placing in an old pillowcase and walking on it until pods are thoroughly crushed. Remove the resulting chaff by pouring the beans back and forth between two pans in a breezy area or in front of a fan. Store in air-tight jars or bags in a dry, cool place.

SPECIAL TIPS: To avoid potential problems with weevils, freeze well-dried beans for several hours before storing.

RELATED PLANTS:

ADZUKI BEAN (*Vigna angularis*): Small, dark red beans of Japanese origin, highly valued for their protein content. 'Express' is an early-maturing cultivar, requiring about 118 frost-free days.

BLACK BEAN (*P. vulgaris*): Sometimes called turtle bean or turtle soup bean, the black bean is especially popular in Caribbean and Latin American cuisines. The beans are small and shiny jet black.

BLACK-EYED BEAN (*Vigna unguiculata* subsp. *unguiculata*): Also called black-eyed pea, this warm-region favorite is related to the asparagus bean, but is grown for its seed rather than its pod. The round, off-white beans are marked with black. It is best in Zones 7 and warmer.

GARBANZO BEAN (*Cicer arietinum*): Also called chickpea, the large, round, buff-colored garbanzo is native to southern Europe and India, where it is

Harvest the nutritious adzuki bean when the pods are completely dry. Store in air-tight jars in a dry, cool place.

Soak garbanzo beans for 2 hours before cooking. You can roast them or serve them as an hors d'oeuvre.

eaten boiled or roasted. It is the main ingredient in hummus. It requires a relatively long growing season, 120 days or more.

HYACINTH BEAN (*Dolichos lablab*): This is a fast-growing, vining plant that is grown in its native tropics for both its pods and its seeds. It is a perennial in warm climates, but can also be grown in cooler areas as an annual. The wisteria-like blossoms are ornamental and the seeds can be dried and eaten like beans.

KIDNEY BEAN (*P. vulgaris*): Historically the term "kidney bean" was used for all common beans, green and shelled as well as dry; but today it is usually used to refer to large, dry beans with the characteristic kidney shape—not just the dark, red-brown ones familiar on the supermarket shelf, but white, brown, yellow, black, and mottled beans as well. Smaller kidney beans are often referred to as navy beans. Among favorite kidney bean cultivars are 'Jacob's Cattle', a red-speckled white bean popular in short-season areas, and 'Soldier', an heirloom bean that derives its name from a soldier-like marking on its edge.

MUNG BEAN (*Vigna radiata*): The seeds are often sprouted for use as bean sprouts, but are also edible as dry beans. The pods may also be eaten when immature.

PINTO BEAN (*P. vulgaris*): The pinto bean is an oval, medium-sized bean, usually mottled on a buff background. Its heat and drought resistance make it a favorite in arid areas. It is the standard bean in Mexican cooking.

Pinto beans are often used in Mexican cooking. You can either use them fresh, or dry the seeds.

Phaseolus vulgaris   Leguminosae

BEAN (FRESH)

Plant wax beans successively for a regular supply. Water frequently when flowers appear for a heavy yield.

You can eat the flowers, pods, and seeds of the versatile scarlet runner bean. In warm climates it is grown as a perennial.

BEST CLIMATE AND SITE: Zones 3 and warmer; cold climates may require a quick-maturing cultivar. Full sun.

IDEAL SOIL CONDITIONS: Likes humusy but not excessively fertile soil; pH 6.0–7.5.

GROWING GUIDELINES: Sow seed after frost danger is past. Sow bush snap beans about 1 inch (2.5 cm) deep and 3 inches (7.5 cm) apart in single or double rows. Keep well weeded, or mulch. Regular watering will increase yield; thorough watering is critical when beans are in flower. Bush snap beans bear heavily but only for a few weeks. To assure a steady supply, make several small plantings 3–4 weeks apart, ending 2 months before the first fall frost date.

PEST AND DISEASE PREVENTION: Mexican bean beetles can be serious pests from midseason onward; early plantings are usually less troubled. Row covers help exclude these pests. Stay out of the bean patch when the plants are wet to avoid spreading bean rust.

COMMON PROBLEMS: Sown too early in cold, wet soil, bean seeds may rot before germinating. A small planting for early harvest may be worth the risk, but larger plantings for freezing or canning should be delayed until the weather is settled.

DAYS TO MATURITY: A minimum of 42–55 frost-free days, depending on species and cultivar.

HARVESTING AND STORING: Pick at any size, but before seeds have begun to swell noticeably inside the pod. Tiny fresh beans are a tender gourmet treat, but larger pods have a more characteristic "beany" flavor. Can, freeze, or pickle snap beans, or leave pods on the plant to mature fully and then dry beans and use as you would any dried bean.

SPECIAL TIPS: Increase yield by using an inoculant of nitrogen-fixing bacteria to help the plant make better use of nitrogen from the air. Add the inoculant to the seed or the soil at planting time.

RELATED PLANTS:

ASPARAGUS BEAN (*Vigna unguiculata* subsp. *sesquipedalis*): Also known as Chinese long bean, dow guak, snake bean, or yardlong bean, this relative of the southern cowpea produces long, thin, crunchy pods that can range up to a yard (almost a meter) long but are better eating at 10–12 inches (about 30 cm). It requires a trellis, likes warm weather, and needs a longer growing season than regular beans—at least 75 days of frost-free weather. Named cultivars are not generally available, but some seed companies offer variants identified by the color of their seed.

BUSH LIMA BEAN (*P. limensis* var. *limenanus*): Sometimes called butter or sugar bean, the lima is a heat-loving bean that does best in Zones 7 and warmer. The seed, rather than the pod, is eaten, either as a fresh vegetable or as a dried bean. The beans are ready to pick when the pods are plump.

Both the seed and pod of butterbeans may be eaten. The yellow pod looks most decorative when served with other vegetables.

Whether eaten fresh or dried, the bush lima bean is a valuable source of nutrients.

BUTTERBEAN (*P. lunatus*): Butterbeans are grown and used like limas but are flat, thin, and much smaller than lima beans. Also called Carolina bean and sometimes, incorrectly, lima bean.

FAVA BEAN (*Vicia faba*): Also called broad bean, the fava is sometimes grown as a shell bean in place of limas in short-season areas. It is far more frost-tolerant than other beans and may be planted with the spring peas, 4–6 weeks before the last frost of spring. It is best in areas with long, cool springs; where winter gives way abruptly to summer, fava blossoms will drop without setting pods. Pick when pods are plump, but before the bean has developed a tough outer skin. 'Ipro' has some tolerance to hot weather; 'Windsor Long Pod' is widely available.

HORTICULTURAL BEAN (*P. vulgaris*): Also called shell bean, shelly bean, or cranberry bean, these plants are grown specifically for their seeds, which are eaten at the immature stage. Harvest horticultural beans when the pods have started to turn rubbery but before the seeds inside have begun to harden. 'Tongue of Fire' is a striking bean, red-streaked on an off-white background; some seed suppliers also carry cultivars of French shelling beans, called "flageolets."

LIMA BEANS (*P. limensis*): Climbing or pole limas require plenty of warm weather and a sturdy trellis, but are considered superior to the bush types in flavor. Requires a minimum 85-day growing season for the first beans to mature; to reap a full harvest will take at least 100 days' favorable weather.

POLE BEAN (*P. vulgaris*): Also called runner beans, these beans require support as they produce long vines. The traditional method is to create a "tee-pee" framework of several long poles tied together at the peak, planting eight to ten seeds at the base of each pole. Pole beans also may be planted in a row and trellised. Pole beans bear over a longer period of time than bush beans and are a good choice where space limitations prevent successive plantings. They require slightly more time to mature, however, and some older cultivars will have strings that must be removed before eating. Cultivars include 'Kentucky Wonder', 'Case Knife', and 'Blue Lake'.

PURPLE-PODDED BEAN (*P. vulgaris*): Striking purple beans that turn dark green when cooked. Quick to mature. Cultivars include 'Royal Burgundy' and 'Royalty Purple Pod'.

ROMANO BEAN (*P. vulgaris*): Sometimes called Italian bean. Tender, well flavored, early to mature. Cultivars include 'Romano II', 'Jumbo', and 'Greencrop'.

SCARLET RUNNER BEAN (*P. coccineus*): A climbing bean that can be grown as a perennial in subtropical areas and as an annual in areas as cool as Zone 4. It has clusters of brilliant red edible flowers, edible

Pole beans are the perfect vegetable for a small garden. They bear over a long period and may be grown on a trellis or fence.

Golden beet, with its attractive golden-orange flesh, adds a gourmet touch to salads. It can be served cooked, too.

pods, and large seeds, which are used as fresh shelled beans. Breeders have created several cool-tolerant cultivars, including 'Prize Winner'.

WAX BEAN (*P. vulgaris*): Also called yellow snap bean. Produces slender, pale to golden yellow pods, which are picked, prepared, and eaten exactly as green snap beans. Cultivars include 'Yellow Pencil Pod' and 'Goldkist'.

BEST CLIMATE AND SITE: Beets thrive in almost any climate, but where summers are hot (Zones 8 and warmer), grow them as a fall, late winter, and early spring crop. Full sun.

IDEAL SOIL CONDITIONS: Well-drained, rich, neutral soil, free of stones. If soil is heavy or shallow, grow only round cultivars; long-rooted ones may be deformed or tough at maturity; pH 6.0–7.5.

GROWING GUIDELINES: Beet seeds are compound—each "seed" actually contains as many as a half-dozen seeds. For this reason, many gardeners sow beets sparingly to reduce thinning tasks. But because beets are known for spotty germination, other gardeners sow them heavily to assure a full stand. The middle course is a moderately heavy seeding after first soaking the seeds in tepid water for several hours to encourage germination. Plant 1 inch (2.5 cm) deep and 2–4 inches (5–10 cm) apart, about 1 month before last spring frost. Firm the seedbed well with your feet or the back of a hoe. Thin the young plants when they are 2–3 inches (5–8 cm) tall. Beets are among the few root vegetables that can be transplanted, so you can move the excess seedlings to another spot, taking care not to double over the taproot when planting. The thinnings may also be used as salad greens. Beets tend to become woody and tasteless when left in the ground too long; small monthly sowings will give you a continuous supply of tender, sweet beets

Romano beans are flat-podded, tender beans, sometimes called Italian beans.

Cylindrical beets are easy to slice into perfect rounds for cooking or canning. They need deeper soil than round beets.

Start harvesting beets when they are 1 inch (2.5 cm) in diameter. You can eat both the leaves and roots.

for the table. The exceptions are cultivars bred for the root cellar, such as 'Lutz Winter Keeper'. Plant these in midsummer, at least 2 months before first fall frost (or in the spring if your growing season is short), and leave in the ground until hard frost threatens. Beets need regular watering to keep them tender and to prevent interior discoloring that results from uneven soil moisture.

PEST AND DISEASE PREVENTION: Row covers will thwart flea beetles and leaf miners. Discourage the disease leaf spot by not growing beets where they or their relatives, such as spinach and chard, have been grown in the previous year.

COMMON PROBLEMS: Beets do poorly in hot weather and dry soil; if your summers are scorchers, grow beets in the cool seasons, including winter in mild-climate areas.

DAYS TO MATURITY: 55–70 days in the garden, or sooner if you grow quick-maturing cultivars such as 'Little Egypt' from seedlings started indoors.

HARVESTING AND STORING: Harvest greens as soon as they are large enough for use. Harvest beets as "babies" when they are 1 inch (2.5 cm) or more in diameter; check by gently probing the soil at the plant's base. When removing the tops, leave 1 inch (2.5 cm) of stem attached to the roots to prevent bleeding. Cook with stems attached and cut away the stems before serving. Pull beets for storage before hard frost, cut tops close to the roots, and

store in sand or sawdust in a cool place. Beets also may be canned or pickled.

SPECIAL TIPS: Grow small-rooted cultivars, such as those bred as specialty "baby beets," in a cluster without thinning. Plant two or three seeds together, each cluster 6 inches (15 cm) apart, and harvest when the beets are eating size.

RELATED PLANTS:

CYLINDRICAL BEET (*B. vulgaris*): Tapered roots, up to 7 inches (18 cm) long and 2 inches (5 cm) in diameter. Hill the crowns to keep the topmost part of root from being exposed and toughened. 'Cylindra' and 'Formanova' are popular cultivars. Easy to peel and slice into uniform pieces for pickling or canning.

GOLDEN BEET (*B. vulgaris*): Sweet and tender-fleshed, but golden orange in color. Prized for its unusual color, which does not "bleed" when cooked as red beets do. Plant golden beets more thickly than other types as they tend to germinate poorly. Cultivars include 'Golden' and 'Burpee's Golden'.

Brassica oleracea, Botrytis group   Cruciferae

BROCCOLI

Virtually unknown in American gardens 50 years ago, broccoli is now a cool-weather favorite.

BEST CLIMATE AND SITE: Zones 3 and warmer; grow as a winter crop in mild-climate areas. Where springs are cool, grow as a spring and fall crop; otherwise does best in fall. Full sun.

IDEAL SOIL CONDITIONS: Well-drained soil with plenty of calcium; pH 6.7–7.2.

GROWING GUIDELINES: Start spring seedlings indoors, about 2 months before the last spring frost. Set out hardened-off transplants in the garden a month before the last frost. Space plants 12–24 inches (30–60 cm) apart; wider spacing will yield larger heads. Cultivate or mulch and keep the soil evenly moist. Lack of water will stress the plant, which may fail to head or may become vulnerable to insect pests. Sow fall crops directly about 90 days before the first fall frost, or transplant about 60 days before frost.

PEST AND DISEASE PREVENTION: Use cardboard or metal "collars" to deter cutworms. Row covers will thwart flea beetles, cabbageworms, and root maggots. Spray or dust with BT if cabbageworms are a problem. To avoid soilborne diseases, don't plant broccoli and other cabbage-family plants in the same spot more than once every 3 years.

COMMON PROBLEMS: In some areas, spring broccoli bolts (goes to seed prematurely) at the onset of hot weather. Choose quick-maturing cultivars and expect spring-grown broccoli heads to be smaller than those grown in fall.

The striking heads of purple broccoli add color to the vegetable garden. It is delicious eaten raw.

DAYS TO MATURITY: 70–95 days.

HARVESTING AND STORING: Harvest broccoli heads when they have reached maximum size but before the tight green flower buds begin to loosen and show yellow. Spring crops are best harvested once, with as much of the edible stem as possible. Cut fall broccoli heads with less stem attached, leaving as much of the plant intact as possible to produce smaller sideshoots or "florets," which you can harvest until hard freeze. Broccoli freezes well.

SPECIAL TIPS: A small planting of Chinese cabbage nearby will draw flea beetles away from broccoli.

RELATED PLANTS:

BROCCOLI RAAB (*B. rapa,* Ruvo group): Also called rapini or rapine; a prized vegetable in Italy that is usually blanched, then chopped and sautéed in olive oil and garlic. The glossy, dark green leaves are harvested along with the flower buds and stems. Spring crops may bolt quickly; fall plantings usually produce three or more harvests of flower shoots and leaves.

CALABRESE BROCCOLI (*B. oleracea,* Italica group): Also called Italian broccoli, asparagus broccoli, or sprouting broccoli. Does not form a large central head but a profusion of smaller, looser heads or "sprouts." Cut when about 4 inches (10 cm) long, with surrounding small leaves, and use like regular broccoli. The plant will produce new sprouts over a long period.

BRUSSELS SPROUTS

Calabrese broccoli is perfect for small gardens. It produces new sprouts over a long period, so you do not need many plants.

CHINESE BROCCOLI (*B. oleracea,* Alboglabra group): Also called Chinese kale, kailaan, or gai lohn. Similar to calabrese, but the buds and leaves are lighter green and sweeter. The stems and young leaves are also eaten steamed or stir-fried. More heat-tolerant than standard broccoli.

PURPLE BROCCOLI (*B. oleracea,* Botrytis group): Identical to green broccoli except that the head has a distinct purplish cast, which disappears during cooking.

ROMANESCO BROCCOLI (*B. oleracea,* Botrytis group): Unusual conical heads of pale green, peaked florets that resemble little rocket ships. The crunchy texture is more like cauliflower than broccoli. Requires wider spacing—up to 36 inches (90 cm)—and a longer growing season than most standard broccoli. In most areas it is planted in late spring for fall harvest.

Individual brussels sprouts resemble tiny cabbages. They are a late-season treat, when frost has sweetened their flavor.

BEST CLIMATE AND SITE: Zones 4–7, and Zone 8 as a winter vegetable. Generally unsuited to warmer climates. Full sun.

IDEAL SOIL CONDITIONS: Well-drained and fertile soil, with adequate calcium levels; pH 6.0–6.8.

GROWING GUIDELINES: Require a long growing season and are best when matured in cool weather. Short-season vegetable gardeners may set out transplants when they sow other spring crops; in other areas, wait until late spring. Space plants about 24 inches (60 cm) apart and keep weeded or mulched. Pinch off top leaves to encourage side growth.

PEST AND DISEASE PREVENTION: Rotate with non-cabbage-family crops to avoid soilborne fungal and viral diseases. Use row covers to deter flea beetles, cabbageworms, and root maggots, or BT for cabbageworms. Keep well watered and grow in fertile soil to reduce vulnerability to aphids.

DAYS TO MATURITY: 90–120 days. Mature plants withstand heavy frost but grow slowly.

HARVESTING AND STORING: Harvest lower sprouts by breaking off the leaf below and snapping off the sprout. Sprouts higher up will continue to grow. Entire stalks can also be harvested. Harvest anytime until early winter to complete harvest before ground freezes. Harvested sprouts freeze well.

SPECIAL TIPS: Sprouts keep for several weeks on the stalk if you pull up the whole plant and keep it in a cold place.

Arctium lappa   Compositae

BURDOCK

Mildly bitter and good in soups and stews, burdock is a nutritious root vegetable.

BEST CLIMATE AND SITE: All Zones, although the eventual length of the root will depend on the growing season. Full sun.

IDEAL SOIL CONDITIONS: Requires deep, well-prepared soil, as the roots can grow to 2 feet (60 cm) or more in length; pH 6.0–7.3.

GROWING GUIDELINES: Sow in spring for fall harvest, or in late summer for overwintering. Thin to 24–36 inches (60–90 cm) apart. Cultivated burdock looks much like its relative common burdock (*A. minus*), with large, heart-shaped leaves, but it can grow to 8 feet (2.4 m). Cultivate shallowly until the plant is large enough to shade out weeds.

PEST AND DISEASE PREVENTION: Burdock is rarely troubled by pests.

COMMON PROBLEMS: In rocky or heavy soil, roots may fork or become deformed, making harvest difficult. Raised beds will help.

DAYS TO MATURITY: 120 days or more.

HARVESTING AND STORING: Harvest after fall frosts, or leave roots in ground for spring harvest. Treated as you would carrots or turnips, burdock stores well in a cool cellar.

OTHER COMMON NAMES: Gobo, nagau pong.

Brassica oleracea, Capitata group   Cruciferae

CABBAGE

Valued for thousands of years because of its hardiness and storage life, cabbage can be grown almost anywhere.

BEST CLIMATE AND SITE: Zones 3 and warmer, with soil-cooling mulch in hotter climates. Use as a winter vegetable in mild areas. Full sun.

IDEAL SOIL CONDITIONS: Any soil texture, provided it is well drained and fertile; pH 6.0–6.8.

GROWING GUIDELINES: Grow as a spring and fall crop in most areas, or as a winter crop where temperatures rarely drop below freezing. Avoid plantings that will mature in hot, dry weather. Start spring crops in a cool place indoors 6–8 weeks before last spring frost and set out 4 weeks before the last frost. Give seedlings plenty of light and withhold fertilizer to discourage spindly growth. Set plants 12–18 inches (30–45 cm) apart, depending on expected head size. In cool-summer areas, you can plant successive crops every month, ending with a storage-type cabbage about 2 months before fall frost. Summer plantings may fail in warmer climates. Heavy mulch will help retain moisture and keep the soil cool. Cabbage prefers rich soil; add compost or rotted manure before planting and apply fish emulsion or compost tea a month after planting.

PEST AND DISEASE PREVENTION: Use row covers or BT to control cabbageworms. Aphids are a sign of heat or water stress; hose them off with a strong water spray or spray plants thoroughly with insecticidal soap. Rotate cabbage-family plantings to avoid soilborne diseases.

Savoy cabbage has puckered leaves that form a looser head than standard cabbage. It is also milder and sweeter.

Red cabbage makes excellent coleslaw and soup, and it is pretty enough to grow in the flower garden.

COMMON PROBLEMS: To avoid head splitting, use a shovel to sever roots on one side of the plant when the head is fully formed. This slows maturity and helps the cabbage "hold" without splitting.

DAYS TO MATURITY: 60–110 days. Young cabbage plants will withstand light frost, and mature ones moderately severe frost.

HARVESTING AND STORING: Harvest early cultivars as needed, as they do not store well. Harvest storage-type cabbage before hard freeze and store, roots and all, in dry leaves in a cold cellar or garage. May be pickled or made into sauerkraut.

SPECIAL TIPS: Cut early and midseason cabbage high on the plant, leaving as many loose, lower leaves as possible. As many as six small cabbages, called "cabbage sprouts," will form on the stem, providing a second harvest.

RELATED PLANTS:

FLOWERING CABBAGE (*B. oleracea,* Acephala group): Actually a kale, rather than a cabbage, flowering cabbage forms a loose head or rosette of colorful green and red, white, or magenta leaves. It differs from flowering kale in that the leaves are smooth-edged rather than frilly or ruffled. Both are edible but are grown chiefly as ornamentals for fall color (cool temperatures are needed to develop the colors). Good cultivars are 'Osaka' hybrids (flowering cabbage) and 'Nagoya' hybrids (flowering kale).

POINTED HEAD CABBAGE (*B. oleracea,* Capitata group): Forms a cone-shaped head. It is quick-maturing, and has been popular for centuries. Cultivars include 'Early Jersey Wakefield', 'Greyhound', and 'Trumpet F1 Hybrid'.

RED CABBAGE (*B. oleracea,* Capitata group): Attractive blue-purple leaves and red-purple heads with white veining. Cultivars include 'Ruby Perfection Hybrid', 'Red Acre', and 'Red Rookie'.

ROUND HEAD CABBAGE (*B. oleracea,* Capitata group): Globe-shaped or flattened round heads. Early cultivars include 'Darkri', 'Polar Green', and 'Emerald Acre'. Midseason cultivars include 'Stonehead' and 'Copenhagen Market'. Late or storage cultivars include 'Late Flat Dutch' and 'Danish Ballhead'.

SAVOY CABBAGE (*B. oleracea,* Capitata group): Puckered, crinkly, light green or blue-green leaves. Heads are often less tight than smooth-leaved cabbage, but flavor is superior. Cultivars include 'Savoy King' and 'Savoy Ace'.

CABBAGE, CHINESE

You can add fast-growing bok choy to stir fries, or even eat it fresh in salads.

Cook choy sum lightly so you can appreciate the flavor of its dark green, glossy leaves.

BEST CLIMATE AND SITE: Zones 2 and warmer. Plant leafy types in summer or early spring in coldest climates. Plant head-forming types in spring and fall in cool climates, fall only in warmer ones; use as a winter vegetable in the warmest climates. Likes full sun.

IDEAL SOIL CONDITIONS: Rich, humusy soil, well drained; pH 6.0–6.8.

GROWING GUIDELINES: Head-forming types of Chinese cabbage quickly go to seed in warm weather or if seedlings are exposed to severe frost, so plant quick-maturing cultivars for early crops. Transplants poorly in spring, but may be moved successfully in peat pots. Direct-seed after last frost; thin to stand 12 inches (30 cm) apart. Set fall crops from transplants or direct-seed 2–3 months before first frost, and give wider spacing, up to 24 inches (60 cm) apart. Transplant or direct-seed leafy types from spring through late summer, thinning to 6–12 inches (15–30 cm) apart, depending on the cultivar.

PEST AND DISEASE PREVENTION: Protect from flea beetles with row covers.

DAYS TO MATURITY: 43–80 days. Mature plants will tolerate moderate frost.

HARVESTING AND STORING: Harvest spring crops before warm weather. Harvest fall cabbages before hard freeze; will keep in a cool cellar for several weeks if trimmed and wrapped in newspaper.

RELATED PLANTS:

BOK CHOY (*B. rapa*, Chinensis group): Also called pak choi, Ching-Chiang cabbage, or Chinese chard. A leafy Chinese cabbage with spoon-shaped, light green to blue-green leaves and thick, crisp stems, favored in stir fry. Cultivars include 'Mei Qing Choi' and 'Joi Choi'.

CHOY SUM (*B. parachinensis*): Similar to bok choy, but the leaves are thick, glossy, and dark green. Named cultivars are not generally available, but seed suppliers may offer variants.

MICHIHLI CABBAGE (*B. rapa*, Pekinensis group): Head-forming cabbage. Tall-growing, cylindrical in shape. Cultivars include 'Jade Pagoda' and 'Michihli'.

NAPA CABBAGE (*B. rapa*, Pekinensis group): Head-forming, barrel-shaped, and larger than Michihli types, up to 10 pounds (5 kg). Early cultivars include 'Springtime' and 'Blues', and late ones include 'China Pride' and 'Wintertime'. 'Lettucy Type' is a Napa cultivar with a loose, open top, like romaine lettuce.

| *Calendula officinalis* | Compositae | *Cynara cardunculus* | Compositae |

CALENDULA

CARDOON

Loved by gardeners for their sunny beauty, calendula's orange and yellow flowers can also spice up salads or soups.

BEST CLIMATE AND SITE: Zones 3 and warmer; in warm climates, grows best in cool seasons. Full sun.

IDEAL SOIL CONDITIONS: Not fussy, but prefers rich, well-drained soil; pH 6.0–7.0.

GROWING GUIDELINES: Direct-seed or transplant 1–2 weeks before last spring frost. Space plants or thin to 6–8 inches (15–20 cm) apart. Mulch to keep the soil cool; water during dry spells.

PEST AND DISEASE PREVENTION: Little troubled, but avoid damage from black blister beetles by not planting near asters.

COMMON PROBLEMS: Calendula is an annual and will die after it sets seed. Pick flowers regularly to extend bloom. When flowering slows, cut entire plant back to 3 inches (7.5 cm) and keep watered for blooms in fall. Or plant a second crop for fall.

DAYS TO MATURITY: 70–90 days. Dwarf cultivars bloom earlier. Will tolerate light frost. May be grown as a pot plant in short-season areas.

HARVESTING AND STORING: Pick fresh blooms as needed; use whole as garnish or pull petals to sprinkle in salads. Dry the petals and store in air-tight jars in a dark place for use as a saffron substitute in rice and soups.

CULTIVARS: 'Pacific Beauty', 'Bon Bon', 'Fiesta Gitana'.

OTHER COMMON NAMES: Pot marigold.

This thistle-like plant, related to the globe artichoke, is grown chiefly for its blanched stalks, which are used much like celery.

BEST CLIMATE AND SITE: Zones 5 and warmer. Perennial in most climates. Where growing seasons are short, cardoon may not reach full height but will still supply usable stalks. Full sun.

IDEAL SOIL CONDITIONS: Well-drained, fertile soil with ample moisture; pH 6.0–7.0.

GROWING GUIDELINES: Start seedlings indoors if season is short; otherwise, direct-seed 1–2 weeks before last frost. Thin to 24 inches (60 cm) apart. A month before fall frost, or earlier if the plant is large enough, tie the leaf stalks together and wrap the plant with burlap or heavy paper to exclude sunlight and whiten the stalks. Stalks are blanched in 3–4 weeks.

PEST AND DISEASE PREVENTION: Few pests. Provide good drainage to prevent crown rot.

DAYS TO MATURITY: 120–150 days for first harvest; in areas where cardoon is perennial, spring shoots may be blanched and eaten like the mature stalks. Will withstand light frost.

HARVESTING AND STORING: Harvest blanched stalks as needed until hard freeze threatens, then bank the plant with earth or straw to prolong harvest into winter. You can also dig and eat the main roots.

CULTIVARS: 'Italian Dwarf', 'Plein Blanc Inerme'.

CARROT

Carrots were well known to the ancient Greeks, who used their delicate foliage in corsages and flower decorations.

Short, plump-rooted carrot cultivars, like 'Early Scarlet Horn', are ideal for growing in containers.

BEST CLIMATE AND SITE: Zones 3 and warmer. Plant quick-maturing carrot cultivars in colder areas. Full sun.

IDEAL SOIL CONDITIONS: Deep, light soil without stones or other obstructions. In heavier or shallow soil, plant round or half-long cultivars. Improve clayey soil with organic matter; pH 5.5–6.8.

GROWING GUIDELINES: Sow the first crop in early spring, after severe frost threats are past. Seed is very small; mix half-and-half with fine sand to help avoid overseeding. Plant a scant ½ inch (12 mm) deep and firm the seedbed gently with the back of a hoe. Mark the row well; carrots are slow germinators and may not appear for 3–4 weeks. Or drop a radish seed every 2 inches (5 cm) into the row with your carrot seeds to help mark it. The quick-germinating radishes will also help break any soil-crusting that could smother the more delicate carrot seedlings, which look like fine blades of grass. Keep the soil evenly moist until the carrots are up. Thin to 2–3 inches (5–8 cm) apart. Plant successive crops every few weeks until 3 months before fall frost. During their growing period carrots should be weeded carefully. Do not overwater as a continually wet soil can cause root rot.

PEST AND DISEASE PREVENTION: Use row covers to deter carrot rust flies, which lay eggs at the base of carrot plants. The larvae feed on the roots. Crops planted after late spring and harvested before late summer often escape damage without protection. Rotate carrot plantings to avoid bacterial diseases.

COMMON PROBLEMS: Twisted roots indicate inadequate thinning. Forked or deformed roots usually mean the seedbed was not fine enough—remove stones, thoroughly break up any clods, and never step on the carrot seedbed. Hairy roots indicate excessive fertility. Do not use high-nitrogen fertilizer or fresh manure on carrot beds. Splitting may occur when heavy rain follows a dry period.

DAYS TO MATURITY: 50–70 days. Will tolerate light frosts as seedlings or as mature plants.

HARVESTING AND STORING: Carrots are ready to eat as soon as they have developed full color. Pull when the ground is moist to avoid breaking roots, or dig with a garden fork. Harvest fall crops before the ground freezes, or earlier if shoulders (root tops) appear aboveground. Dry in the sun and store, packed in dry leaves or straw, in a cool, moist place. Carrots may be harvested while they are still young. This will spread the harvest, and the "baby" carrots are delicious either raw or cooked.

SPECIAL TIPS: Cover newly seeded carrot rows with boards or black plastic for 3 weeks, then remove. This will preserve soil moisture and prevent sprouting of weeds, which can easily get ahead of the much slower germinating carrots. In gardens where space is limited, grow carrots in the ornamental garden. Their brilliant green feathery tops make a

CAULIFLOWER

The long roots of Nantes or Amsterdam type cultivars are best suited to moderately deep soils.

Cauliflowers are heavy feeders that require plenty of water. Blanching is necessary to keep the heads white.

pretty addition to a border. Small-rooted carrot cultivars adapt readily to pot culture.

CULTIVARS: 'Parmex' and 'Planet' are early-maturing, short, round carrots, good for heavy soils or short seasons. Cultivars such as 'Scarlet Nantes' and 'Minicor' are suitable for moderately deep soils. Long carrot cultivars for very deep and light soils include 'Imperator' and 'Gold Pak'.

BEST CLIMATE AND SITE: Zones 3 and warmer. Use as spring and fall crops in cool-spring areas, and as a winter vegetable in mild climates. Full sun.

IDEAL SOIL CONDITIONS: Rich, well-drained soil with plenty of calcium, and ample amounts of well-rotted manure or compost worked in; pH 6.0–7.0.

GROWING GUIDELINES: Start spring crops indoors and do not set out too early; severe frost may cause the plant to form a "button" instead of a full-sized head. Space plants 18–36 inches (45–90 cm) apart, and mulch to keep soil cool. Start fall crops 90–120 days before fall frost; set out when seedlings are 4–5 weeks old. Keep cauliflower growing steadily with plenty of water and one or more applications of fish emulsion or compost tea. When heads appear, use a clothespin to clip several large leaves together over the head to shade it and keep it white, or remove a large lower leaf and lay it over the developing head. "Self-blanching" cultivars grow their own "shading" leaves, but may still need blanching if heat wilts protective leaves.

PEST AND DISEASE PREVENTION: Control cabbageworms with row covers or a spray or dust of BT. Do this early, as it is difficult to control worms inside the developing head. Row covers will also deter flea beetles on young plants; older, vigorous plants can withstand flea beetle attack. Rotate cabbage-family plantings to avoid soilborne diseases.

Purple cauliflower is easier to grow than white cultivars. Its mild, broccoli-like head requires no blanching.

The yellow-green head of the cultivar 'Alverda' looks particularly attractive in the garden and on the plate.

COMMON PROBLEMS: In humid areas, heads may discolor during blanching due to excess moisture. Clip leaves together loosely to shade the head without cutting off the air circulation.

DAYS TO MATURITY: 50–72 days. 'Snow Crown' and 'Rushmore' are early cultivars; longer-season cultivars include 'Snowball Elite' and 'Andes'. Young plants will withstand light frost; mature ones, moderate frost.

HARVESTING AND STORING: Cut heads before curds begin to coarsen and separate, usually after 5–7 days of blanching. Cool weather may slow head development. When harvesting, leave a few leaves around the head to avoid breaking it. Cauliflower keeps refrigerated for up to 2 weeks and may be frozen.

RELATED PLANTS:

GREEN CAULIFLOWER (*B. oleracea*, Botrytis group): A specialty cauliflower with yellow-green, rather than white, curds. Green cauliflower, like purple cauliflower, is easier to grow than the white cultivars, since it does not require blanching. Cultivars include 'Alverda'.

PURPLE CAULIFLOWER (*B. oleracea*, Botrytis group): Like a cauliflower in growth habit, but has looser flower heads, like broccoli. Purple cauliflower is milder and more tender than broccoli. The head has a deep purple cast, which disappears during cooking. No blanching is required. Cultivars include

'Purple Head Improved', 'Violet Queen', and 'Burgundy Queen'. The attractive purple flower heads may be used as a garnish or to add color to salads.

The heads of 'Purple Head Improved' add color to salads. The color disappears during cooking.

CELERIAC

In salads, soups, and stews, or as a cooked vegetable, the root of celeriac has all the flavor of celery.

BEST CLIMATE AND SITE: Zones 5 and warmer. Use as a fall and winter crop in mild areas. Full sun.

IDEAL SOIL CONDITIONS: Rich, moisture-retentive soil, with adequate calcium and plenty of well-rotted manure or compost worked in; pH 5.5–7.5.

GROWING GUIDELINES: Start indoors 6–8 weeks before last spring frost and set out, 10–12 inches (25–30 cm) apart, when the frost threat is past. Celeriac may be direct-seeded, but it germinates slowly and can be overtaken by weeds. Keep bed well weeded and watered. Apply compost tea or fish emulsion at least once a month.

PEST AND DISEASE PREVENTION: Rotate plantings of celeriac and celery to avoid blights to which they are both vulnerable. Handpick celeryworms or parsleyworms (green caterpillars with yellow and black bands).

COMMON PROBLEMS: Inadequate moisture and fertility produce small, tough, or fibrous celeriac.

DAYS TO MATURITY: 110–120 days; can withstand increasingly severe frost for the last 30–45 days.

HARVESTING AND STORING: Harvest the turnip-like root when large enough for your needs. Harvest all plants before the ground freezes; cut stems close to the roots and store like turnips in damp sawdust or sand in a cool place.

CULTIVARS: 'Large Smooth Prague' widely available.

OTHER COMMON NAMES: Celery root, German celery, knob celery, turnip celery.

CELERY

It isn't easy to grow, but with a little attention to its needs, celery does beautifully in the backyard.

BEST CLIMATE AND SITE: Zones 5 and warmer, except where early summer temperatures fall below 55°F (13°C). Fall and winter crop in mild areas. Full sun.

IDEAL SOIL CONDITIONS: Moisture-retentive, rich soil, with adequate calcium and plenty of well-rotted manure or compost worked in; pH 5.5–7.5.

GROWING GUIDELINES: Start indoors 4–6 weeks before last spring frost. Germination and seedling growth are slow. Keep seedlings watered and do not expose to temperatures below 55°F (13°C), which can cause plants to bolt (go to seed). Set plants out when the weather is well settled (about a month after the last frost), 10–12 inches (25–30 cm) apart. Do not allow the soil to dry out; feed plants with fish emulsion or compost tea at least once a month. If desired, you can blanch celery before harvest by slipping a bottomless paper bag over the plant and tying it in place, or by putting wide boards on edge on both sides of the celery row and holding the boards in place with stakes.

PEST AND DISEASE PREVENTION: Rotate plantings to avoid blight problems. Handpick parsleyworms or celeryworms (green caterpillars with yellow and black bands).

COMMON PROBLEMS: Excessive heat, inadequate moisture, or lack of fertility will result in tough or stringy celery.

DAYS TO MATURITY: 80–105 days. Young plants cannot tolerate frequent temperatures below 55°F

Lactuca sativa var. *asparagina*   Compositae

CELTUCE

The stems of 'Improved Utah' celery give a continuous harvest for 2 to 3 months if picked as needed like Swiss chard.

(13°C); mature plants withstand severe frost.

HARVESTING AND STORING: Cut celery for immediate use just below soil level. Before hard freeze, pull entire plant and roots and store, packed in dry leaves or straw, in a cool cellar or garage.

SPECIAL TIPS: Plant celery in well-manured beds, three or four plants abreast and 10 inches (25 cm) apart. Dense growth will shade out weeds and automatically blanch the celery.

CULTIVARS: 'Improved Utah' and 'Tendercrisp', both green cultivars. 'Stokes Golden Plume', a yellow celery.

Celtuce has lettuce-like leaves and celery-like stalks. Cooked, it tastes like a cross between summer squash and artichoke.

BEST CLIMATE AND SITE: Zones 3 and warmer. In colder areas, harvest leaves only, if the stalk takes too long to develop. Full sun.

IDEAL SOIL CONDITIONS: Well-drained soil; otherwise not fussy; pH 6.0–6.7.

GROWING GUIDELINES: Sow in early spring and again in midsummer. Best if matured in cool weather. Seeds are small; sow thinly and cover lightly. Thin to 12 inches (30 cm) apart, adding thinnings to salad.

PEST AND DISEASE PREVENTION: Usually free from pests and diseases.

COMMON PROBLEMS: Water frequently to delay bittering of leaves and help keep the central stalk tender. Warm weather induces bolting, which toughens the stalk.

DAYS TO MATURITY: 45 days for leaves (thinnings may be harvested earlier); 90 for stalks. Tolerates light frost.

HARVESTING AND STORING: Pick leaves as needed. Use young leaves in salad, and cook older ones. Cut the stalk at its base and peel for use as a raw or boiled vegetable, or in stir fries and soup.

CULTIVARS: Named cultivars not generally available.

OTHER COMMON NAMES: Asparagus lettuce.

Sechium edule Cucurbitaceae	*Cichorium intybus* Compositae
# CHAYOTE	# CHICORY

Chayote is a space-hungry avid climber suited to warm climates. Shoots, roots, and fruits are all edible.

BEST CLIMATE AND SITE: Zones 7 and warmer. Cooler areas may produce fruits only; roots are not harvested until the second year. Full sun, adjacent to a wall or fence for climbing.

IDEAL SOIL CONDITIONS: Deep, fertile, moist but well-drained soil; pH 5.5–6.5.

GROWING GUIDELINES: After last spring frost, plant entire fruit (about 4 inches/10 cm long containing one seed), with 1 inch (2.5 cm) of narrow end exposed. One plant will satisfy most gardeners' needs. Plant near a fence or trellis, or provide 12 feet (3.6 m) of growing space. Keep well watered. Perennial in mild areas; may overwinter in cooler climates with protection. Cut back after frost and mulch roots; pull back mulch after frost danger is past next spring.

PEST AND DISEASE PREVENTION: Provide good air circulation to avoid fungal disease and mildew.

DAYS TO MATURITY: 80–90 days for fruits; shoots and roots are harvested the second year.

HARVESTING AND STORING: Pick immature fruits at 4–6 inches (10–15 cm) and use like summer squash. Harvest mature fruits before frost; store like winter squash and use in stews. Cut second-year shoots in spring; marinate or steam as you would asparagus. Dig roots in fall of the second year; store and use them as you would potatoes.

OTHER COMMON NAMES: Cho-cho, choko, christophene, chuchu, mirliton, vegetable pear.

The leaves of cutting chicory impart a tangy taste to green salads. Cut the leaves from the plant as needed.

BEST CLIMATE AND SITE: Zones 3 and warmer; only leafy types grow in coldest areas, and they also grow as a winter crop in mild areas. Some cultivars are suited to fall planting for spring harvest in areas as cool as Zone 5. Full sun.

IDEAL SOIL CONDITIONS: Well-drained, moderately fertile soil; pH 6.0–6.5.

GROWING GUIDELINES: Direct-sow after frost or set out as transplants, 8–12 inches (20–30 cm) apart. Sow or transplant the fall crop 2 months before first frost. Avoid plantings that will mature during hot weather. Keep well watered to avoid bolting. Modern cultivars of head-forming chicories do not require cutting back to form heads.

PEST AND DISEASE PREVENTION: Chicory is little troubled by pests.

COMMON PROBLEMS: Keep the soil moist to avoid interior browning of head-forming types. Head-forming types are inconsistent; many plants will produce only leaves or loose rosettes despite the best conditions.

DAYS TO MATURITY: 40–130 days, depending on type. Leafy types, such as cutting chicory, are fast-growing. Witloofs require a long season; head-forming types are intermediate. Chicories vary in frost resistance; most will tolerate only light frost.

HARVESTING AND STORING: Harvest leaves of leafy types as needed. Cut head-forming types as needed when the heads are firm (plants will hold in the

Heading chicory, also known as radicchio, is popular. In Europe, it is often served broiled or roasted as well as in salads.

'Elodie', a cutting chicory with frilled leaves, makes a pretty border to the vegetable or flower garden.

garden for several weeks). Harvest chicories grown for their roots before hard frost.

CULTIVARS:

CUTTING CHICORY (*C. intybus*): Also called catalogna, leaf chicory, or Italian dandelion. It is grown for its leaves, which vary in shape from dandelion-like to smooth or rosette-like, and in color from light to dark green and also red-tinged. Some may be cut and left to resprout for continued harvest. Cultivars include 'Catalogna', 'Elodie', 'Rouge di Verone', and 'Bionda'.

FORCING CHICORY (*C. intybus*): Also called witloof and Belgian endive. Cut initial head in fall, then dig roots, trim to 8 inches (20 cm), and bury upright in moist sand or peat. Kept in complete darkness and at 50–60°F (10–15°C) temperatures, roots will produce small pale sprouts or "chicons" in about 3 weeks. Cultivars include 'Witloof Improved'.

HEADING CHICORY (*C. intybus*): Also called radicchio or Italian chicory. Red-tinged leaves surround small, red head with white veining, resembling red cabbage. Older cultivars, such as 'Rossa di Treviso', may require cutting back to 1 inch (2.5 cm) in midsummer to form heads in fall. Modern cultivars form heads without cutting back. Cultivars include 'Adria', 'Giulio', and 'Rossa di Verona'. 'Pan di Zucchero', or sugar loaf, makes large, light green heads similar to romaine lettuce.

ROOT CHICORY (*C. intybus*): Also called coffee chicory or Sicilian chicory. Has edible leaves, but is usually grown for its large roots, which you can dig in fall, dry, and grind to make a coffee-like beverage. Cultivars include 'Magdeburgh'.

Chrysanthemum coronarium   Compositae

CHRYSANTHEMUM, GARLAND

The aromatic edible leaves of the garland chrysanthemum are popular in oriental cuisines, either stir-fried or in soups.

BEST CLIMATE AND SITE: Zones 2 and warmer; use as a winter vegetable in mild areas. Avoid hot, dry areas.

IDEAL SOIL CONDITIONS: Not fussy, but prefers fertile, moist soil; pH 6.0–6.7.

GROWING GUIDELINES: Direct-seed about ½ inch (12 mm) deep in early spring or fall and thin to 4–6 inches (10–15 cm), using thinnings in stir fry, soup, or salad.

PEST AND DISEASE PREVENTION: Little troubled by pests, but to avoid leaf diseases, don't plant where garden chrysanthemums have been grown.

COMMON PROBLEMS: Hot weather and inadequate moisture can cause premature bolting. The flowers, however, are attractive.

DAYS TO MATURITY: 45 days, but thinnings may be harvested earlier. Tolerates light to moderate frost.

HARVESTING AND STORING: Harvest whole plants or side leaves as needed, but avoid plantings in hot weather, as the leaves become disagreeably bitter. Garland chrysanthemum "cooks down" in the pot, so several plants may be needed for a soup.

SPECIAL TIPS: Remove flower buds to prolong harvest of leaves.

CULTIVARS: No named cultivars, but seed companies may offer variants based on leaf size and shape.

OTHER COMMON NAMES: Crown daisy, edible chrysanthemum, shungiku, tong ho choi.

Brassica oleracea, Acephala group   Cruciferae

COLLARD

A cabbage relative, collard is among the most nutritious of garden greens. It can tolerate heat but is tastier after frost.

BEST CLIMATE AND SITE: Zones 3 and warmer; use as a winter vegetable in mild areas. Full sun.

IDEAL SOIL CONDITIONS: Fertile, moist soil with adequate calcium; pH 6.5–6.8.

GROWING GUIDELINES: Sow in early spring and thin to 18 inches (45 cm), or transplant as you would cabbage. Spring-planted collards, if well cared for, will produce past frost, but you can also plant in midsummer for fall crop only. Tolerates summer heat, but prefers cool weather, so mulch to keep the soil cool. Keep moist and apply fish emulsion or compost tea once a month for lush growth.

PEST AND DISEASE PREVENTION: Protect young plants from flea beetles and cabbageworms with row covers. Rotate with other cabbage-family members to avoid soilborne diseases.

COMMON PROBLEMS: Direct-seeding too thickly may cause overcrowded and spindly plants. Thin gradually—first to 6 inches (15 cm), then 12 inches (30 cm), then 18 inches (45 cm)—for early greens for the table.

DAYS TO MATURITY: 65–80 days, although you can harvest thinnings earlier. Young plants withstand light frost; mature ones, severe frost.

HARVESTING AND STORING: Harvest leaves as needed from bottom of plant; leave the crown to sprout new leaves. Harvest crown before hard freeze for tender "baby collards." Freeze for winter use.

CULTIVARS: 'Champion', 'Georgia Blue Stem', 'Vates'.

Zea mays var. *rugosa*   Gramineae

CORN

Tender and succulent, fresh sweet corn is a classic summer treat from the garden. It's well worth growing.

The colorful kernels of ornamental corn make an interesting kitchen decoration, and the plant is very easy to grow.

BEST CLIMATE AND SITE: Zones 3 and warmer. Use only short-season cultivars in coldest areas. Full sun. Avoid high-wind areas.

IDEAL SOIL CONDITIONS: Accepts many soil textures; prefers deep, well-manured soil with high organic content; pH 6.0–6.8.

GROWING GUIDELINES: Sow after last spring frost, or 1–2 weeks earlier if soil has first been warmed by covering with black plastic for a week or more in sunny weather. Corn seed germinates poorly in cold, wet soil and may rot. Plant 1 inch (2.5 cm) deep and 4 inches (10 cm) apart in blocks at least four rows wide, rather than in long, single rows, to ensure good wind pollination. Rows should be 18–24 inches (45–60 cm) apart. Thin seedlings to stand 8–12 inches (20–30 cm) apart; mulch or cultivate shallowly to avoid damaging roots near the soil surface.

Corn grows rapidly and needs adequate fertility and water. Apply fish emulsion or compost tea after 1 month and again when the tassels appear. Water is most critical when corn is in tassel.

Plant successive crops every 10–14 days through midsummer, or·until about 90 days before first frost. Choose an early-maturing cultivar for the last planting. Modern corn cultivars include some known as "super-sweet," which have been genetically modified to slow the conversion of sugar to starch. If these cultivars cross-pollinate with other sweet corn cultivars, though, tough kernels will result. Separating corn plots by 25 feet (7.5 m) or more is recommended, but not always practicable in home gardens. Instead, time the plantings to ensure that at least 10 days elapse between pollination periods for the cultivars. For example, you could plant a super-sweet cultivar that matures in 82 days at the same time as a regular cultivar that matures in 72 days. Or you could plant an 82-day super-sweet cultivar adjacent to an 82-day regular cultivar as long as you sow it 10 days earlier or later.

PEST AND DISEASE PREVENTION: Plant in warm soil to avoid wireworms, which destroy seed. Late plantings, at spring's end, are less prone to corn borers; early plantings are less susceptible to corn earworms. Help control earworms by dropping mineral oil into the immature ears as soon as you see the pest's sticky frass on the silks (the hair-like fibers). Rotate corn plantings and shred or bury crop debris to reduce overwintering pests.

COMMON PROBLEMS: Patchy spots on kernels, or ears that are not filled to the tip, indicate inadequate pollination. In small plots, hand-pollinate by stripping pollen from the tassels and sprinkling it on the silks, especially on plants on outside edges of the plot.

DAYS TO MATURITY: 54–94 frost-free days. Early cultivars (54–70 days) include 'Seneca Horizon' and 'Earlivee'. Longer-season cultivars include

Children love growing and cooking popcorn. Store the dried kernels in air-tight jars for future use.

You can hand-pollinate corn, and increase your crop, by stripping pollen from the tassels and sprinkling it on the silks.

'Tuxedo', 'Honey 'n Pearl' (a super-sweet cultivar), and 'Silver Queen'.

HARVESTING AND STORING: Harvest when the silks have turned brown and dry and the ear feels full, usually about 3 weeks after the silks appear. Check by pulling back a husk and pressing a thumbnail into a kernel. It should squirt back a milky liquid. Harvest ears by holding the stalk firmly and snapping the ear downward, then up. Quickly can or freeze corn that cannot be eaten immediately after picking, as it begins to turn starchy within hours.

SPECIAL TIPS: In gardens where space is limited, interplant a fast-growing crop like lettuce between the rows of corn. The lettuce is harvested before the corn casts too much shade.

RELATED PLANTS:

MINIATURE CORN (*Z. mays* var. *rugosa*): Also called baby corn. Pick tiny ears a day or two after silks appear and use whole in stir fries or for pickling. 'Baby' and 'Baby Asian' are grown specifically for miniature corn, but any sweet corn cultivar will do.

ORNAMENTAL CORN (*Z. mays* var. *indurata*): Also called Indian corn. These are colorful cultivars of field corn, with kernels of many hues. Many cultivars, such as 'Hopi Blue' and 'Mandan Bride', are excellent for grinding as cornmeal or flour. 'Indian Fingers' is a miniature ornamental corn, 3–5 inches (8–13 cm) long. You should harvest when the kernels cannot be penetrated with a thumbnail, then pull back the husks and hang to finish drying.

POPCORN (*Z. mays* var. *praecox*): The dense kernels explode into crisp puffs when heated. Harvest as dry as possible, pull back the husks, and hang in an airy place to "cure." Test-pop a few kernels to determine when the corn is dry enough to be shucked and stored in bags or jars. Cultivars include 'Matinee', 'Strawberry' (decorative corn with poppable kernels), and 'White Cloud'.

| *Valerianella locusta* | Valerianaceae | *Lepidium sativum* | Cruciferae |

CORN SALAD

CRESS

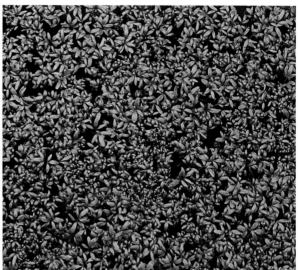

Small, tender, green corn salad can be planted in fall to grace spring's earliest salads. It's expensive to buy, but easy to grow.

Cress is quick and easy to grow. It adds grace to the table in salads or as a garnish.

BEST CLIMATE AND SITE: Zones 2 and warmer. Overwinters in Zones 5 and warmer; grow as a winter vegetable in mild areas. Full sun or partial shade.

IDEAL SOIL CONDITIONS: Not fussy, but prefers rich, humusy soil; pH 6.0–7.0.

GROWING GUIDELINES: Sow in early spring, 2–4 weeks before last frost; just cover with fine soil and firm with the back of a hoe. Plant thickly in rows or more thinly in wide beds; keep moist. Thin to stand 2 inches (5 cm) apart; use thinnings in salads. Avoid plantings that will mature in hot weather. Plant in fall near first frost date; mulch lightly after hard freeze. Remove mulch in early spring.

PEST AND DISEASE PREVENTION: Little troubled by pests or diseases.

COMMON PROBLEMS: In colder areas, mulch more heavily to avoid heaving of plants over winter.

DAYS TO MATURITY: 45–60 days from spring seeding; thinnings can be harvested earlier. Will withstand light frost.

HARVESTING AND STORING: Harvest entire rosettes by pinching off at ground level. Some cultivars remain sweet even when in flower; taste to check.

SPECIAL TIPS: In wide-bed plantings, thin and weed simultaneously by raking shallowly in two directions.

CULTIVARS: 'Elan', 'Piedmont', 'Vit'.

OTHER COMMON NAMES: Fetticus, lamb's lettuce, mache.

BEST CLIMATE AND SITE: Zones 2 and warmer; grow as a winter vegetable in mild areas. Full sun or partial shade.

IDEAL SOIL CONDITIONS: Rich, moist, humusy soil; pH 6.0–7.0.

GROWING GUIDELINES: Sow thickly in rows in early spring, 4–6 weeks before last frost. Barely cover and firm the soil with back of a hoe. Sow successive small plantings every 7 days until 2 weeks after frost. Later plantings may bolt in summer heat. Sow fall crop 2–4 weeks before first frost.

PEST AND DISEASE PREVENTION: Do not grow where cabbage-family members have grown the previous year. Row covers will deter flea beetles.

COMMON PROBLEMS: Hot weather and inadequate water will cause cress to bolt prematurely.

DAYS TO MATURITY: 21-35 days. Will withstand severe frost.

HARVESTING AND STORING: Cut with shears just above the crown and let plants resprout for two or three more cuttings.

RELATED PLANTS: Upland cress (*Barbarea verna*). Also called American cress, scurvy grass, or creasy greens. Has a spicy flavor similar to watercress. Forms a small rosette of dark green, rounded leaves, which may be harvested whole or leaf by leaf. Sow as above; thin to 6 inches (15 cm).

CULTIVARS: Named cultivars not generally available.

OTHER COMMON NAMES: Curly cress, peppergrass.

Cucumis sativus Cucurbitaceae

CUCUMBER

Prized in salads and pickles, cucumbers are heat-loving plants that can be grown on a trellis or fence to save garden space.

The burpless cucumber was especially developed to eliminate the substance in the skin that makes cucumbers difficult to digest.

BEST CLIMATE AND SITE: Zones 4 and warmer. Full sun, in a site with good air circulation.

IDEAL SOIL CONDITIONS: Light soil with well-rotted manure or compost worked in; pH 6.0–7.0.

GROWING GUIDELINES: Cucumbers dislike cold soil, so wait for 3–4 weeks after the last frost to direct-seed or transplant them into the garden. Indoors, sow seeds in peat pots 2–3 weeks before you plan to set them out, or presprout in a moist paper towel and plant into soil that has been warmed by covering with a sheet of black plastic mulch. Thin plants to stand 12–15 inches (30–38 cm) apart in rows, or plant in "hills" (each with four to five plants) about 4 feet (1.2 m) apart. Mulch, or keep cultivated until plants begin to vine. Cucumbers need ample food and water; irrigate in dry spells and give them fish emulsion or compost tea once a month. Plant successive crops 6 weeks apart, where the growing season permits. Grow on a trellis or fence to save space.

PEST AND DISEASE PREVENTION: Use row covers to protect young plants from cucumber beetles, but remove when plants bloom. Trellising will improve air circulation and reduce mildew. Do not grow cucumbers or their relatives, such as squash and melons, in the same spot more often than once every 3 years.

COMMON PROBLEMS: Sudden collapse of cucumber plants indicates wilt disease, which is spread by cucumber beetles. Straw mulch may help deter beetle attack, and some cultivars are less vulnerable to the beetles. Successive plantings will extend the harvest even where beetles are troublesome.

DAYS TO MATURITY: 48–70 frost-free days, slightly less for gherkins or baby pickles.

HARVESTING AND STORING: Cucumbers are best picked small: 3–4 inches (8–10 cm) for pickling types and 6–8 inches (15–20 cm) for slicers. A yellowish tinge at the blossom end indicates overmaturity. Pick often, especially during hot spells, to encourage continued fruiting. Pickle excess cucumbers.

RELATED PLANTS:

'BURPLESS' CUCUMBER (*C. sativus*): Also called Lebanese cucumber or oriental cucumber. Cultivars include 'Sweet Slice', 'Burpless Hybrid', and 'Japanese Long Pickling'.

GYNOECIOUS CUCUMBER (*C. sativus*): Bred to produce only female flowers, which greatly increases yield per plant. Gynoecious cucumbers require another cucumber cultivar to pollinate them (seeds of a pollinator are included with packets of gynoecious cucumbers). They also need higher soil fertility to support higher yields. Cultivars include 'Saladin Hybrid' and 'Early Pride'.

PICKLING CUCUMBER (*C. sativus*): Also called gherkins or cornichons. Although all cucumbers can be pickled, some cultivars were selected for this

CUCUMBER— Continued

DANDELION

Gynoecious cucumbers thrive in rich soil and a sunny position. They produce a heavy crop over a long period.

Often scorned as a common weed, dandelion is useful in salads and as a pot herb. The flowers are used for wines and jellies.

purpose because of their small size, crispness, or small seed cavity. Often lighter green than slicing cultivars, with soft "spines" on the fruit. Cultivars include 'Picklebush', 'Vert de Massy', and 'Bush Baby'.

SERPENT CUCUMBER (*Trichosanthes anguina*): Also called snake gourd or club gourd. A popular vegetable in Asian countries. The thin, long fruits will curve and coil on the ground, but grow straight if the vine is trellised.

SLICING CUCUMBER (*C. sativus*): These cultivars are generally grown for fresh or salad use, rather than pickling, and are usually longer, smoother, and darker green than the picklers. Cultivars include 'Straight Eight', 'Spacemaster', and 'Marketmore'.

BEST CLIMATE AND SITE: Zones 3 and warmer. Full sun.

IDEAL SOIL CONDITIONS: Not fussy; best in deep, light soil for the largest roots; pH 5.8–6.2.

GROWING GUIDELINES: Sow in early spring, covering lightly. Thin to 10 inches (25 cm) apart, and add thinnings to salad. Keep spent flowers pinched off to encourage leaf growth and avoid spreading seed to places where plant is not desired. In colder areas, mulch lightly after a hard freeze, as cultivated strains often are not as hardy as wild ones.

PEST AND DISEASE PREVENTION: Little troubled by pests and diseases.

COMMON PROBLEMS: If leaves are excessively bitter, blanch the leaves by inverting a clay pot over the plant for a week or more before harvest.

DAYS TO MATURITY: 120 days, although thinnings may be harvested earlier. Dandelion is a perennial and will provide early spring greens starting in the second year.

HARVESTING AND STORING: Pick flowers for wine and jelly in early morning, before petals begin to dry. Harvest leaves as needed, blanching first if desired. Dig roots of second-year or older plants, dry, and grind to make a coffee-like beverage. Roots also may be forced like chicory (see "Forcing Chicory" on page 102) for pale, mild greens.

CULTIVARS: Named cultivars not generally available, but seed suppliers may offer cultivated strains.

Hemerocallis spp. Liliaceae	*Solanum melongena* var. *esculentum* Solanaceae

DAYLILY

EGGPLANT

Daylily buds and flowers may be eaten, either stuffed or in salads. Raw and cooked roots are often used in Asian food.

Eggplant thrives in hot weather. It needs ample amounts of water and fertilizer to produce.

BEST CLIMATE AND SITE: Zones 3 and warmer. Full sun; but will tolerate partial shade.

IDEAL SOIL CONDITIONS: Well-drained soil; otherwise not fussy; pH 5.5–7.0.

GROWING GUIDELINES: Some daylilies grow from seed, but it is quicker to plant tuberous roots. Set roots with the crown at soil level, 12–24 inches (30–60 cm) apart. Closely spaced, daylilies quickly fill the bed, eliminating weed problems, but you will have to divide sooner. Dress bed each year with compost, but do not overfertilize, or leaves will be lush and flowers skimpy. Divide after bloom, midsummer to fall, by digging up clumps and cutting into smaller pieces with a knife or cleaver.

PEST AND DISEASE PREVENTION: Little troubled.

COMMON PROBLEMS: Scant blooms suggest overly rich soil or a daylily bed in need of dividing.

DAYS TO MATURITY: Spring-planted daylilies may bloom the first year; otherwise, the second year.

HARVESTING AND STORING: Pick buds and flowers as needed. As the name implies, daylily blossoms last but one day. Dig tuberous roots anytime from established plantings, or harvest at dividing time and store in peat moss in a cool, dry place.

SPECIAL TIPS: Remove pollen-bearing stamens from daylilies before eating if you suffer from hayfever. Sample the roots in a small quantity first. Adverse reactions to daylily roots are rare but have occurred.

CULTIVARS: There are many named cultivars.

BEST CLIMATE AND SITE: Zones 5 and warmer. Full sun.

IDEAL SOIL CONDITIONS: Light, rich, well-drained, warm soil; pH 5.5–6.8.

GROWING GUIDELINES: Eggplant likes hot weather, so it does not pay to start the plants too soon or set them out too early. Start 8 weeks before the time when night temperatures can be counted on to stay above 50°F (10°C). Night covers may be needed in colder climates. Space the plants 18–24 inches (45–60 cm) apart. Eggplant needs plenty of water and food to produce well. Use fish emulsion or compost tea at least once a month.

PEST AND DISEASE PREVENTION: Use row covers to deter flea beetles. Handpick Colorado potato beetles; a light dusting of ground limestone often helps repel this pest. Rotate eggplant and other members of the nightshade (Solanaceae) family, such as tomatoes and potatoes, to avoid soilborne diseases such as Fusarium and Verticillium wilt. If the soil is already infected, grow eggplant in large pots in disease-free soil.

COMMON PROBLEMS: Provide adequate soil moisture and calcium to prevent blossom-end rot, which shows up as a soft, brown spot at the blossom end of the fruit.

DAYS TO MATURITY: 50–75 frost-free days.

HARVESTING AND STORING: Clip the fruit with some stem attached; harvest often to encourage

In standard black-purple, or a rainbow of other colors, eggplant is as decorative as it is delicious.

Eggplant is an essential ingredient in Italian cooking. One 'Black Beauty' plant will produce four to ten large fruits.

further fruiting. Standard eggplant cultivars, such as 'Black Beauty', will bear four to ten full-sized fruits, or more if picked small. Oriental cultivars, such as 'Sanshi', bear more but smaller fruits. It's better to harvest when a little small than to let eggplant become overmature and bitter. When frost threatens, cut all fruits, leaving 2 inches (5 cm) of stem attached, and store in a cool, moist, well-ventilated area, where they will keep for a month or longer. Eggplant does not freeze or can well.

SPECIAL TIPS: Where the growing season permits, two small plantings a month apart will provide a steadier supply of fruit.

RELATED PLANTS:

BABY EGGPLANT (*S. melongena* var. *esculentum*): A gourmet item, usually slender or egg-shaped and 4 inches (10 cm) long or less. Quick-maturing, good for short-season areas. Cultivars include 'Pirouette' and 'Little Fingers'.

ITALIAN EGGPLANT (*S. melongena* var. *esculentum*): Italians love eggplant and have developed many fine cultivars with tender, nonbitter white flesh and dazzling skin color. 'Rosa Bianca' is rosy lavender and white; 'Violetta di Firenze' has dark lavender, somewhat ridged fruit.

ORIENTAL EGGPLANT (*S. melongena* var. *esculentum*): Also called Japanese eggplant. Asians favor long, thin fruits, usually purple-black but also light green and white-lavender. Cultivars include 'Orient Express' and 'Kurume'. Oriental seed suppliers often offer eggplant variants by fruit description.

WHITE EGGPLANT (*S. melongena* var. *esculentum*): White-skinned eggplant, attractive but often less prolific than its dark purple siblings. 'Casper' is a full-sized cultivar; 'Osterei' and 'White Egg' bear smaller fruits.

Cichorium endiva   Compositae

ENDIVE AND ESCAROLE

In warm areas grow escarole in between tomatoes, where it will appreciate the shade and save space.

Add the curly leaves of endive to an ordinary salad plate to give it a gourmet touch. The leaves have a tangy taste.

BEST CLIMATE AND SITE: Zones 4 and warmer; grow as a winter vegetable in mild areas. Full sun or partial shade in warm areas.

IDEAL SOIL CONDITIONS: Fertile, well-drained, and well-limed soil; pH 5.5–7.0.

GROWING GUIDELINES: In cool-spring areas, start indoors 8 weeks before last frost and set out a month before last frost, 10–12 inches (25–30 cm) apart. Cover plants if hard frost threatens. In warmer areas, they're best as fall crops, started in midsummer and set out 2 months before first fall frost, or even later in mildest areas. May be blanched by inverting a pot or box over the plant, or by tying outer leaves together at the top. In humid areas, blanching by tying may cause inside leaves to rot.

PEST AND DISEASE PREVENTION: Provide good air circulation to avoid molds. Control slugs with shallow pans of beer set into the soil.

COMMON PROBLEMS: Water well and provide shade in hot spells to prevent brown, crisp ends on endive leaves.

DAYS TO MATURITY: 80–100 days.

HARVESTING AND STORING: Cut the entire head at its base in the morning while dew is still on the plant. Rinse; wrap in a paper towel and then in a plastic bag. Keeps up to 2 weeks refrigerated.

SPECIAL TIPS: For information on Belgian endive, see "Forcing Chicory" on page 102.

CULTIVARS: 'Full Heart Batavian' is a widely available cultivar of escarole, the name used for broadleaf endive. Curly- or fringed-leaved cultivars of endive, also called frisée, include 'Salad King', 'Giant Fringed Oyster', and 'Traviata'.

Foeniculum vulgare var. *azoricum* Umbelliferae

FENNEL, FLORENCE

Florence fennel is grown chiefly for its enlarged leaf bases, which have the flavor of licorice.

BEST CLIMATE AND SITE: Zones 4 and warmer; grow as a winter vegetable in mild areas. Full sun.

IDEAL SOIL CONDITIONS: Fertile and well-drained soil; pH 6.5–7.0.

GROWING GUIDELINES: In cool-spring areas, direct-seed spring crop 2–4 weeks before last frost and thin to stand 8–10 inches (20–25 cm) apart. Transplants poorly, but may need a head start in warmer areas since it bolts quickly in hot weather. Start seedlings in individual pots or peat pots a month before last frost; do not let roots outgrow the pots before setting out. Sow small, successive crops where the season permits; avoid plantings that will mature in hot, dry weather. Fall crops are easier to transplant; set out 2 months before first frost. Keep well weeded or mulch; water in dry spells.

PEST AND DISEASE PREVENTION: Handpick celery- or parsleyworms, which are green caterpillars with black and yellow stripes.

COMMON PROBLEMS: Florence fennel becomes woody when overmature. Plant small, successive crops to prolong harvest.

DAYS TO MATURITY: 65–90 days. Tolerates light frost.

HARVESTING AND STORING: Cut the entire plant below the base of the bulb. Use fronds as seasoning for fish. Fennel does not freeze well.

SPECIAL TIPS: Use ripe seeds of bolted plants.

CULTIVARS: 'Zefa Fino', 'Romy'.

OTHER COMMON NAMES: Finocchio.

Allium sativum Liliaceae

GARLIC

Garlic is an easy-to-grow and rewarding crop. Plant it in fall to harvest next summer.

BEST CLIMATE AND SITE: Zones 3 and warmer; must be prechilled before planting to form bulbs in warmer climates. Full sun or partial shade, but bulbs will be smaller in partial shade.

IDEAL SOIL CONDITIONS: Well-drained, fertile, humusy soil; pH 6.0–7.0.

GROWING GUIDELINES: Best planted in fall, 2–4 weeks before first frost. Plant individual cloves, pointed end up, 1–2 inches (2.5–5 cm) deep and 4–6 inches (10–15 cm) apart. Deeper planting is best where frequent freezing and thawing may cause heaving (when the soil moves due to temperature changes and plants are dislodged from their roots). Apply fish emulsion or compost tea in spring to encourage vigorous growth. Keep the bed cultivated or mulched. Irrigate until tops begin to brown, then withhold water to let the plant dry naturally.

PEST AND DISEASE PREVENTION: Little troubled by pests, but reduce threat of root maggots by not planting where garlic or its relatives, such as onions and shallots, have been planted the previous year.

COMMON PROBLEMS: Humid conditions at harvest may induce neck rot. Use these heads immediately or freeze, as they will not store.

DAYS TO MATURITY: 120–150 days for spring plantings; fall plantings are harvested the following summer.

HARVESTING AND STORING: Use spring shoots and

GOURD, BOTTLE

Elephant garlic is technically a leek, but its very large heads have a mild garlic flavor.

Though edible, bottle gourds are most commonly grown as ornamentals. Hang dried, hollowed-out gourds for birdhouses.

flower stems chopped in salads and dips, as you would chives. Pull mature garlic plants when about 75 percent of the foliage is brown; tie in bundles and dry in a dark, well-ventilated place. Braid for storage or clip stems and store in mesh bags in a dry, cool area. Freeze peeled cloves; thaw slowly before using.

SPECIAL TIPS: Pot garlic heads or cloves and grow indoors for wintertime use as greens.

RELATED PLANTS:

ELEPHANT GARLIC (*A. ampeloprasum*, Ampeloprasm group): Elephant garlic is popular for its very large heads, which can weigh up to a pound (500 g), and for its less pungent cloves. Grow like garlic, but plant deeper—up to 4 inches (10 cm) deep.

HARD-NECK GARLIC (*A. sativum* var. *ophioscorodon*): Also called top-setting garlic, serpent garlic, or rocambole. Produces a coiled flower stem, prized in oriental cuisine. The stem must be removed to direct growth to the roots. Cloves are large and form in a single ring around a central, woody stem. Cultivars include 'Russian Red' and 'Chet's'.

SOFT-NECK GARLIC (*A. sativum* var. *sativum*): The standard commercial type, also called artichoke garlic. Cloves form in an overlapping pattern, like the scales of an artichoke. More productive than hard-neck types, but the cloves are smaller. The soft stem makes it easy to braid. Cultivars include 'French Silverskin' and 'California Early'.

BEST CLIMATE AND SITE: Zones 5 and warmer. Full sun, good air circulation.

IDEAL SOIL CONDITIONS: Well-drained, rich soil; pH 6.0–6.5.

GROWING GUIDELINES: Plant after frost danger is past. Sow two or three seeds to a hill and site hills 6 feet (1.8 m) apart, or sow more closely if you will train the vine to a trellis or fence. Mulch or keep weeded. If allowed to sprawl, the vining plant eventually will shade out weeds. It is drought-tolerant, but irrigate for larger and more numerous fruits.

PEST AND DISEASE PREVENTION: Use a trellis to improve air circulation and lessen mildew problems. Rotate plantings of gourds and their relatives, such as squash and melons.

COMMON PROBLEMS: In short-season areas, pinch off flowers after the plant has set several fruits, to ensure maturity before frost.

DAYS TO MATURITY: 90 frost-free days for mature gourds; fewer for immature, edible-sized fruits.

HARVESTING AND STORING: Pick young fruits and prepare as you would summer squash. For use as a birdhouse, dry the mature gourd until seeds rattle inside. Cut it by drilling many small holes close together, and wax the gourd to preserve it.

CULTIVARS: Types of *L. siceraria* include calabash, dipper gourd, and sugar trough.

OTHER COMMON NAMES: Cucuzzi, po gua, yugao.

GROUND CHERRY

HORSERADISH

Favored for pies and preserves, this plant produces loads of yellow cherry-sized fruit, faintly tomato-flavored.

BEST CLIMATE AND SITE: Zones 4 and warmer. Warm, sheltered spot in cold areas.

IDEAL SOIL CONDITIONS: Not fussy, but prefers rich, well-drained soil; pH 6.0–6.5.

GROWING GUIDELINES: Start indoors 4–6 weeks before last spring frost; set out 12–18 inches (30–45 cm) apart after frost danger is past. In short-season areas, cover if fall frost threatens before fruit is ripe.

PEST AND DISEASE PREVENTION: Little troubled by pests and diseases.

COMMON PROBLEMS: May drop blossoms without setting fruit in hot spells. Keep watered; it will continue to bloom.

DAYS TO MATURITY: 70 frost-free days.

HARVESTING AND STORING: Pick fruit when papery husk is dry and berry inside is plump and well colored. Each plant will produce about a pint (500 ml) of berries. Stew them with sugar.

RELATED PLANTS:
TOMATILLO (*P. ixocarpa*): Also called Mexican ground cherry. Green to light yellow fruit in husks, used in salsas and chilies. A prolific, sprawling plant. Space plants 24–36 inches (60–90 cm) apart. Named cultivars are not generally available, but seed companies sometimes offer an ornamental purple-leaved variant.

OTHER COMMON NAMES: Dwarf cape gooseberry, husk tomato, strawberry tomato.

Not actually a radish, horseradish is a hardy perennial grown for its roots, which are grated for a tangy condiment.

BEST CLIMATE AND SITE: All Zones, but best in Zones 5 and warmer. Full sun.

IDEAL SOIL CONDITIONS: Well-drained, deep, and preferably light soil; pH 6.0–7.0.

GROWING GUIDELINES: Plant root divisions in spring, 2 inches (5 cm) deep and 12–18 inches (30–45 cm) apart. Weed or mulch after leaves appear above the ground. Even soil moisture encourages rapid growth and large roots.

PEST AND DISEASE PREVENTION: Flea beetles sometimes attack horseradish but rarely do more than cosmetic damage.

COMMON PROBLEMS: Avoid growing in heavy, shallow, or stony soil, to prevent horseradish roots from being small, deformed, and difficult to grate. May be difficult to move or eradicate once established, as new plants grow from bits of root left in the soil.

DAYS TO MATURITY: 150–180 days in first year where the season permits; otherwise first harvest is in the second year. Severe frost will kill horseradish leaves but will not damage roots.

HARVESTING AND STORING: Dig roots when they are large enough for your needs. Well-grown ones will be fairly smooth, 12–15 inches (30–38 cm) long and up to 2 inches (5 cm) thick. Dig as needed until the ground freezes. Peel root and grate it by hand or in a food processor. Add salt and vinegar to taste; keeps in the refrigerator for up to 6 weeks, or may be canned. Store whole roots in sand or

JERUSALEM ARTICHOKE

The pretty white flowers of horseradish add an ornamental touch to the vegetable garden. Use them as a garnish.

Jerusalem artichoke is a hardy perennial vegetable. It produces tasty potato-like tubers.

sawdust in a cool place. The young leaves of horseradish can be added to salads.

SPECIAL TIPS: Pack smaller roots in sawdust and save for planting next year. Cut tops flat and bottoms at an angle, so they go in the ground right-side up, producing straighter and more uniform roots. To control the spread of this invasive vegetable, plant it in a bottomless bucket that you've sunk into the soil.

CULTIVARS: Named cultivars not generally available.

BEST CLIMATE AND SITE: All Zones. Full sun or partial shade. Plant away from other garden beds, as most cultivars are perennial in all Zones and many are invasive.

IDEAL SOIL CONDITIONS: Not fussy, but prefers loose, fertile soil; pH 6.0–6.5.

GROWING GUIDELINES: Plant pieces of tuber 6–8 weeks before last spring frost or in fall. Plant like potatoes, each piece with one or more "eyes," 3 inches (7.5 cm) deep and 12–18 inches (30–45 cm) apart in rows or beds. Mulch or weed early; plants will grow quickly to shade out weeds.

PEST AND DISEASE PREVENTION: Rarely troubled by pests and diseases.

COMMON PROBLEMS: Difficult to eradicate once established. Thorough harvesting each year will keep the bed under control.

DAYS TO MATURITY: 110–150 days for a spring-planted, first-year crop. Fall-planted crops are ready next fall. Plants withstand moderate frost; freezing does not damage tubers.

HARVESTING AND STORING: Mulch bed to keep ground diggable after hard freeze. They do not store well, so dig as needed. Prepare them as you would potatoes.

CULTIVARS: 'Fuseau', a smooth-skinned yam type; 'French Mammoth White', traditional knobby tuber; 'Sugarball', round.

OTHER COMMON NAMES: Sunchoke.

Brassica oleracea, Acephala group   Cruciferae

KALE, CURLY

Hardy and nutritious, curly kale provides tasty greens late into the season, even under a blanket of snow.

BEST CLIMATE AND SITE: All Zones; may overwinter in areas as cool as Zone 6. Full sun.

IDEAL SOIL CONDITIONS: Fertile and well-drained soil, with adequate calcium; pH 6.5–6.8.

GROWING GUIDELINES: Direct-seed 4–6 weeks before last spring frost, about ½ inch (12 mm) deep. Thin to 12–15 inches (30–38 cm) apart; use thinnings in salads. Cultivate shallowly or mulch. Tolerates heat but prefers cool soil. Plant fall crop 6–8 weeks before first frost, or later to overwinter. Give fish emulsion or compost tea liberally once a month to encourage lush leaf growth.

PEST AND DISEASE PREVENTION: Use row covers to protect young plants from flea beetles and cabbageworms, or use BT to control cabbageworms. Rotate kale and other cabbage-family plants to avoid soilborne diseases.

COMMON PROBLEMS: Leaves toughen with age. Keep plant well picked; plant a fall crop for a second harvest of tender leaves. Sweeter after frost.

DAYS TO MATURITY: 55–65 days, although thinnings may be harvested sooner. Young plants will take light frost; mature ones, severe frost.

HARVESTING AND STORING: Pick leaves as needed from base of plant; new ones will sprout from top.

CULTIVARS: 'Dwarf Blue Curled Vates'; 'Red Russian', a noncurly cultivar with purplish stems and leaves; 'Winterbor'.

OTHER COMMON NAMES: Borecole, colewort.

Crambe maritima   Cruciferae

KALE, SEA

Grow sea kale for its spring shoots, which you can blanch and eat like asparagus.

BEST CLIMATE AND SITE: Zones 4 and warmer. Full sun or partial shade.

IDEAL SOIL CONDITIONS: Rich, well-drained, slightly alkaline soil; pH 6.0–7.0.

GROWING GUIDELINES: Best started from divisions, but usually only seeds are available, from specialty suppliers. Plant about six seeds in hills 24–36 inches (60–90 cm) apart, 2–4 weeks before last spring frost. Thin to three or four plants per hill. In colder areas, mulch over winter. Top-dress with compost or well-rotted manure each spring.

PEST AND DISEASE PREVENTION: Rarely troubled by pests.

COMMON PROBLEMS: Sea kale roots tend to rise to the surface. Cover exposed roots with soil each fall to prevent winter-kill and increase shoot production the following year.

DAYS TO MATURITY: Grown from seed, sea kale will not provide a harvest until the third year. After that, harvest annually in the spring.

HARVESTING AND STORING: In late winter, before the first shoots appear, invert a large clay pot or basket over the hill. To speed shoot emergence, mulch around the pot with straw or dry leaves. Check periodically for shoots, and cut them with a knife below the soil surface when they are 6–8 inches (15–20 cm) long. Cut third-year plants for 1–2 weeks; harvest older plants for 3–4 weeks. Remove the pot when harvest is finished.

Brassica oleracea, Gongylodes group   Cruciferae

KOHLRABI

Both the swollen, turnip-like stem and the leaves of kohlrabi are edible, and they are often cooked together.

BEST CLIMATE AND SITE: Zones 3 and warmer. Grow in spring and fall in most areas, and as a winter vegetable in mild areas. Full sun.

IDEAL SOIL CONDITIONS: Fertile and well-drained soil with adequate calcium; pH 6.0–7.0.

GROWING GUIDELINES: Direct-seed 2–4 weeks before last spring frost or start indoors 6–8 weeks before last frost and set out 6 inches (15 cm) apart. Use thinnings of direct-seeded crops as salad greens. Cultivate or mulch. Set fall crops from transplants, or direct-seed 2 months before fall frost.

PEST AND DISEASE PREVENTION: Use row covers to protect young plants from flea beetles and cabbageworms, or control cabbageworms with BT. Rotate kohlrabi and other cabbage-family plants to avoid soilborne diseases.

COMMON PROBLEMS: Keep watered or mulched and avoid plantings that will mature in hot weather to prevent tough, woody kohlrabi.

DAYS TO MATURITY: 38–55 days. Will withstand light frost.

HARVESTING AND STORING: Harvest entire plant when swollen stem is about 2 inches (5 cm) in diameter. Overgrown kohlrabi may be woody. Best eaten fresh, but will keep for 2 weeks in the refrigerator.

CULTIVARS: 'Early Purple Vienna' (purple-skinned cultivar), 'Early White Vienna', 'Grand Duke Hybrid'.

Allium ampeloprasum, Porrum group   Liliaceae

LEEK

Leeks are grown for their stout, flavorful stem. These onion relatives hold well in the ground for late harvest.

BEST CLIMATE AND SITE: Zones 3 and warmer; grow as a winter vegetable in mild areas. Full sun.

IDEAL SOIL CONDITIONS: Loose, very rich, well-drained soil; pH 6.0–7.5.

GROWING GUIDELINES: Start indoors up to 12 weeks before last spring frost. Transplant from seed flats to small, individual pots when large enough to handle. This produces larger transplants and better leeks. Set out after frost, 6 inches (15 cm) apart, in a 6-inch (15 cm) deep trench or in holes made with a hoe handle or dibble, covering all but 1 inch or so (2–3 cm) of leaves. Keep well weeded. As leeks grow, fill in the trench gradually or, if planted on level soil, "hill" them by drawing soil up around the stems. This produces a longer white stem, which is the edible part. You can also use a deep mulch to help blanch stems. Keep the soil moist, especially early in the season.

PEST AND DISEASE PREVENTION: To avoid root maggot damage, do not plant where other onion family members have grown the previous year.

COMMON PROBLEMS: Short, tough stems indicate lack of moisture or fertility or inadequate hilling.

DAYS TO MATURITY: 70–105 days in garden. 'King Richard' and 'Titan' are early leeks; 'Carina' and 'Alaska' are good for overwintering. "Baby" leeks may be harvested earlier. Young plants take light frost; mature ones, severe frost.

HARVESTING AND STORING: Dig or pull when large

LEEK—CONTINUED

LENTIL

The leaves of Asian leek add a delicious aroma to soups and stews. The flowers make a pretty garnish.

A staple in Mediterranean cuisines, the protein-rich red lentil is a relative of the pea and is nearly as easy to grow.

enough for use. Before hard freeze, mulch heavily to keep the bed diggable in winter. Harvest overwintered leeks before spring growth begins. Packed in damp sand or peat moss, harvested leeks will keep 6–8 weeks in a cool place.

RELATED PLANTS: ASIAN LEEK (*A. tuberosum*): Also called Chinese leek, Chinese chives, or garlic chives. A hardy perennial grown for its mildly garlic-flavored leaves, which are used in stir fries and soups. The spring flowers also are edible. Named cultivars are not generally available.

CULTIVARS: 'Carina' (large, thick stems, matures in late fall, very cold-tolerant); 'Splendid' (long, thin, tender stems, produces high yields); 'Titan' (long, thick stems, grows quickly).

BEST CLIMATE AND SITE: Zones 5 and warmer. Full sun.

IDEAL SOIL CONDITIONS: Light, well-drained soil; pH 5.8–6.2.

GROWING GUIDELINES: Sow 1–2 weeks before last spring frost, about 1 inch (2.5 cm) deep and 3–4 inches (8–10 cm) apart. Plants have tendrils, like peas, but grow only about 16 inches (40 cm) tall and do not need a trellis. Weed or mulch and irrigate during dry spells until the pods begin to dry, then withhold water to hasten drying.

PEST AND DISEASE PREVENTION: Avoid mildew by providing good air circulation. If weevils are a problem, do not grow where beans or peas have grown in the previous year.

COMMON PROBLEMS: Rainy or humid weather may cause mature seeds to sprout or rot in pods. Pull plants when most of the foliage has died; hang by roots in a well-ventilated place to finish drying.

DAYS TO MATURITY: 120–150 days.

HARVESTING AND STORING: Harvest as you would dry beans, but do not remove from pods until ready to use; lentils keep better in the pod. Freeze well-dried pods for several hours to avoid potential weevil problems, and store in air-tight cans or jars.

CULTIVARS: Named cultivars not generally available.

Lactuca sativa   Compositae

LETTUCE

Leaf lettuce is the quickest and easiest-to-grow member of the lettuce clan. It can double as an ornamental garden edging.

Frequent watering is the secret to success when growing 'Buttercrunch' lettuce. It may also be grown in containers.

BEST CLIMATE AND SITE: All Zones; grow as a winter vegetable in mild areas. Full sun in cool weather, partial shade in warm weather.

IDEAL SOIL CONDITIONS: Fertile, well-drained soil; pH 6.0–6.8.

GROWING GUIDELINES: Direct-seed leafy types a month before last frost. Barely cover the seeds and firm the soil with the back of a hoe. Start head-forming types indoors 6–8 weeks before last frost and set out, 8–10 inches (20–25 cm) apart, 1–2 weeks before last frost. Leave leaf lettuces unthinned for "baby" salad greens or thin to encourage large-leaved plants. Sow or transplant successive crops until a month before first frost, ending with a cold-tolerant cultivar such as 'Winter Density'. Grow summer-maturing crops in partial shade, or cover with a lath frame to shield them from midday sun. Cultivate shallowly or mulch; water frequently, especially during hot spells.

PEST AND DISEASE PREVENTION: Quick-growing lettuce seldom has pest problems, but rotate crops to avoid soilborne diseases. Control slugs by trapping them in shallow pans of beer set flush with the soil surface.

COMMON PROBLEMS: Bitter leaves suggest heat and water stress, or simply overmature lettuce. Plant small crops at frequent intervals.

DAYS TO MATURITY: 40–90 days. Young leaves and thinnings may be harvested earlier. Leaf lettuces mature most quickly, followed by bibb and romaine. Head lettuces require a longer season. Plants will withstand light frost or moderate frost with protection.

HARVESTING AND STORING: Harvest as needed for fresh use. Cut unthinned leaf lettuce with scissors 1 inch (2.5 cm) above roots, leaving the plants to resprout for a second harvest. Cut larger whole plants at soil level.

SPECIAL TIPS: Lettuce germinates poorly at temperatures above 80°F (27°C). Prechill seed before sowing, or start summer crops from transplants.

RELATED PLANTS:

BIBB LETTUCE (*L. sativa*): Also called Boston, butterhead, or cabbage lettuce. Forms loose heads of soft, folded leaves. Cultivars include 'Buttercrunch', 'Dark Green Boston', 'Sangria' (red-tinged leaves), and 'Tom Thumb' (miniature bibb).

ICEBERG LETTUCE (*L. sativa*): Also called crisphead or cabbage lettuce. Forms a tight head of crisp-textured leaves. Prefers a long, cool growing season. Cultivars include 'Great Lakes' and 'Ithaca'.

LEAF LETTUCE (*L. sativa*): Also called cutting lettuce. Instead of heads, it forms loose leaves that may be harvested while quite small. Cultivars include 'Black Seeded Simpson', 'Red Sails', 'Salad Bowl', and 'Oak Leaf'.

ROMAINE LETTUCE (*L. sativa*): Also called Cos lettuce. Forms an upright, elongated head of crisp,

Luffa acutangula and *L. aegyptiaca*   Cucurbitaceae

LUFFA

More tolerant of heat than most lettuces, romaine is a good choice for summer crops. The upright leaves have a crisp texture.

ribbed leaves. Has a better heat tolerance than does iceberg. Cultivars include 'Parris Island', 'Valmaine', and 'Rosalita' (red-leaved cultivar).

Ideal for areas with long, cool springs, iceberg lettuce forms a cabbage-like head of crisp, juicy leaves.

Luffa produces squash-like fruits that are edible when young. Its mature fruits are used as sponges.

BEST CLIMATE AND SITE: Zones 5 and warmer. Full sun, near a fence or trellis.

IDEAL SOIL CONDITIONS: Rich, well-drained soil; pH 6.0–6.5.

GROWING GUIDELINES: In warm areas, plant after last frost, 2–3 seeds in hills about 6 feet (1.8 m) apart, or closer if a trellis is available. In colder areas, start indoors 2–3 weeks before last frost and set out when frost danger is past. Train to a fence or trellis; luffa vines will grow to 15 feet (4.5 m) or more.

PEST AND DISEASE PREVENTION: Little troubled by pests and disease, but provide good air circulation.

COMMON PROBLEMS: In cooler areas, pinch off flowers after several large fruit have set, to ensure maturity before frost.

DAYS TO MATURITY: 80 days for immature, edible fruits. Up to 150 days for fully mature fruits.

HARVESTING AND STORING: Pick fruits when 4–6 inches (10–15 cm) long and prepare them as you would summer squash. For sponges, leave fruits on until they yellow, then dry in a well-ventilated place. Soak to loosen the skin from the fibrous interior, remove skin, and let dry again. Shake out seeds.

CULTIVARS: Named cultivars not generally available. *L. acutangula* produces 15-inch (38 cm), dark green, smooth-skinned fruit; *L. aegyptiaca* produces 20-inch (50 cm), light green, ridged fruit.

OTHER COMMON NAMES: Chinese okra, dish cloth vine, loofah, vegetable sponge.

MELON

Heat-loving muskmelons often require a little extra care in the garden, but the vine-ripened fruit is a great treat at harvest.

Casaba melons mature later in the season than other melons. They also keep for a long period.

BEST CLIMATE AND SITE: Zones 4 and warmer, using short-season cultivars in colder areas. Full sun in a site with good air circulation.

IDEAL SOIL CONDITIONS: Well-manured, well-drained soil, not too acid; pH 6.0–7.0.

GROWING GUIDELINES: In long-season areas, direct-seed after frost and when the soil is thoroughly warm. Plant three or four seeds to a hill and site hills 4–5 feet (1.2–1.5 m) apart. In colder areas, start indoors in individual pots, two seeds to a pot, around the time of last frost. Set plants out in 2–3 weeks into soil that has first been warmed with black plastic mulch. Use row covers to increase warmth and protect from cucumber beetles; remove covers when the plants bloom. Cultivate shallowly until plants begin to vine. Water weekly with compost tea or fish emulsion to keep the vines growing strongly.

PEST AND DISEASE PREVENTION: Plant radishes or basil in the hill with melons to help deter cucumber beetles. Or protect plants with row covers; remove covers when plants start to flower. Spray foliage thoroughly with a baking soda solution (1 teaspoon per quart [liter] of water) to control mildew. Rotate plantings of melons and relatives, such as squash and cucumber, to reduce pest populations.

COMMON PROBLEMS: Wilt spread by cucumber beetles causes the plant to collapse before ripening fruit. Keep plants growing strongly; do not allow them to become stressed from lack of water or nutrients. Where the season permits, successive plantings increase the odds of a good harvest.

DAYS TO MATURITY: 68–95 days. Cantaloupes and muskmelons usually mature earlier, watermelons and honeydews later. Will not tolerate frost.

HARVESTING AND STORING: Pick muskmelons when cracks appear where fruit is attached to the stem. They should "slip" from the stem easily with a slight tug. Let them sit 1–2 days at room temperature to ripen perfectly. Watermelons are ripe when the tendril closest to the fruit is brown and shriveled. Ripening of hard-rind melons, such as honeydews, casabas, and true cantaloupes, is usually signaled by a subtle change in rind color. Most melons have a short storage life; some watermelons and late-season melons will keep up to 2 months in a cool, dry place.

RELATED PLANTS:

APPLE MELON (*C. melo,* Chito group): Also called mango melon, vegetable orange, garden lemon, or vine peach. Small yellow or orange fruit, used in preserves and pickles. No named cultivars.

BITTER MELON (*Momordica charantia*): Also called balsam pear, la-kwa, foo gwa, or bitter cucumber. Oblong or oval orange-yellow warty fruit with a bitter flavor, used in Asian cuisines. Requires a trellis.

True cantaloupes do not have netting on the skin. They are fragrant, sweet, and very popular in Europe.

Despite its name, watermelon is more drought-resistant than many melons, and less prone to disease as well.

CANTALOUPE (*C. melo,* Cantalupensis group): Also called rock melon. Most commercial "cantaloupes" are really muskmelons. True cantaloupes, widely grown in Europe, do not have netted rinds. Cultivars include 'Charentais'.

CASABA MELON (*C. melo,* Inodorus group): Large, football- or banana-shaped fruit with gold skin and white flesh. Late-maturing, good keeper. Cultivars include 'Juan Canary' and 'Santa Claus'.

CRENSHAW MELON (*C. melo,* Inodorus group): Pear-shaped with pale tan skin, delicately flavored light pink or salmon flesh. Cultivars include 'Burpee Early Hybrid Crenshaw'.

CITRON MELON (*Citrullus lanatus* var. *citroides*): Also called preserving melon. Small melon with hard, white flesh. Use preserved.

HONEYDEW MELON (*Cucumis melo,* Inodorus group): Smooth rind, usually pale green or ivory, with green flesh. Generally late-maturing, although there are early cultivars. Cultivars include 'Earlidew' and 'Venus Hybrid'.

MUSKMELON (*C. melo,* Reticulatus group): Also called Persian melon, nutmeg melon, and, incorrectly, cantaloupe. The standard commercial melon. Rind is netted and usually ridged, flesh is orange or green, sweet and fragrant. Cultivars include 'Harper Hybrid' (small, early melon), 'Ambrosia', and 'Iroquois'.

PICKLING MELON (*C. melo,* Conomon group): Also called oriental melon. Oblong, with crisp flesh. Used for pickling in Asian cuisines.

POMEGRANATE MELON (*C. melo,* Dudaim group): Also called Queen Anne's pocket melon, or stink melon. Small, flattened fruit, with marbled or striped rind. Grown chiefly for its sweet melon fragrance, not for its thin, flavorless flesh.

SERPENT MELON (*C. melo,* Flexuosus group): Slender, coiled fruit up to 3 feet (90 cm) long. Sometimes pickled, but grown mostly as a novelty.

WATERMELON (*Citrullus lanatus*): More drought-tolerant than other melons and less vulnerable to insect pests. Rind can be dark to light green, solid, striped, or mottled. Usually red-fleshed, but there are orange-, yellow-, and white-fleshed cultivars. Cultivars include 'Sugar Baby' (small "icebox-sized" melon), 'Charleston Gray' (large, long season), and 'Yellow Doll' (yellow-fleshed).

WINTER MELON (*Benincasa hispida*): Also called wax gourd, hairy gourd, Chinese watermelon, zit-kwa, tunka, or tong qwa. Large fruit, up to 50 pounds (25 kg), with white flesh. Popular in Asian countries for preserves, pickles, and soups; occasionally eaten raw.

Brassica juncea var. *japonica*　　　　Cruciferae

MIZUNA

Milder than most mustards and easy to grow, mizuna adds dash to salads and stir fries, or can be cooked as greens.

BEST CLIMATE AND SITE: All Zones; grow as a winter vegetable in mild climates. Full sun.

IDEAL SOIL CONDITIONS: Not fussy, but best in light, well-drained soil; pH 5.8–6.2.

GROWING GUIDELINES: Sow 2–4 weeks before last spring frost, or 6–8 weeks before first fall frost as a fall crop. Thin to stand 4–6 inches (10–15 cm) apart, using thinnings in salad. Irrigate to promote rapid growth.

PEST AND DISEASE PREVENTION: Row covers will protect young plants, especially spring crops, from flea beetles.

COMMON PROBLEMS: Mizuna can become strong-flavored in hot weather. Avoid plantings that will mature during hot spells, or use as cooked greens then, rather than in salads.

DAYS TO MATURITY: 40 days; thinnings may be harvested earlier. Will withstand moderate frost.

HARVESTING AND STORING: Leaves will grow to 12–14 inches (30–35 cm); cut when 4–6 inches (10–15 cm) long for best salad greens. Harvest as needed with scissors about 1 inch (2.5 cm) above the crown; the plant will sprout new leaves. Four or five harvests are possible. Or harvest whole plants for cooking greens.

CULTIVARS: 'Tokyo Beau', 'Tokyo Belle'.

OTHER COMMON NAMES: Japanese mustard, kyona.

Brassica spp.　　　　Cruciferae

MUSTARD

Black mustard is grown mainly for its seeds, which produce a spicy condiment. The leaves are delicious in salads.

BEST CLIMATE AND SITE: All Zones; grow as a winter vegetable in mild areas. Full sun.

IDEAL SOIL CONDITIONS: Not fussy, but prefers light, well-drained soil; pH 5.8–6.2.

GROWING GUIDELINES: Sow 2–4 weeks before last spring frost or 6–8 weeks before first fall frost. Thin to stand 4–6 inches (10–15 cm) apart, adding thinnings to salads or soups. Irrigate to promote rapid growth.

PEST AND DISEASE PREVENTION: Rarely troubled by pests. Good air circulation prevents mildew.

COMMON PROBLEMS: Goes to seed quickly in hot weather. Avoid plantings that will mature in hot, dry weather; grow as a fall crop in warm-spring areas.

DAYS TO MATURITY: 40–50 days; thinnings may be harvested earlier. Will withstand moderate frost.

HARVESTING AND STORING: Pick leaves as needed from the outside of the plant, allowing new leaves to sprout from the center. Tender leaves do not store well in the refrigerator.

RELATED PLANTS:

BLACK MUSTARD (*B. nigra*): Edible leaves, but grown chiefly for its seed, which is ground to produce a spicy condiment.

BROWN MUSTARD (*B. juncea* var. *crispifolia*): Large, curly or frilled green leaves, usually grown for greens. Cultivars include 'Green Wave' and 'Southern Curled'.

The leaves of brown mustard are valued for salad greens. Use them in combination with cress for a tangy salad.

RED MUSTARD (*B. juncea* var. *foliosa*): Broad, flat leaves with red or purple coloring, very spicy. Cultivars include 'Osaka Purple' and 'Red Giant'. WHITE MUSTARD (*B. hirta*): Seeds are ground for condiment; less pungent than black mustard.

The spicy leaves of red mustard add color to a green salad. Use them sparingly.

Abelmoschus esculentus   Malvaceae

OKRA (GUMBO)

The essential ingredient in gumbo, okra is also admired for its lovely, hollyhock-like flowers. It loves hot weather.

BEST CLIMATE AND SITE: Zones 5 and warmer. Full sun.

IDEAL SOIL CONDITIONS: Fertile and well-drained soil; pH 6.5–7.5.

GROWING GUIDELINES: Sow when frost danger is past and the soil is warm. Prewarming the soil with black plastic mulch will speed germination. In cooler areas, start indoors in individual pots 2–4 weeks before last frost and set out when the weather is settled. Set out or thin to stand 12–15 inches (30–38 cm) apart. Grows quickly in warm weather and needs both food and water; irrigate in dry spells and give it compost tea or fish emulsion once a month.

PEST AND DISEASE PREVENTION: Little troubled by pests. Aphids are a sign of water stress.

COMMON PROBLEMS: In cooler areas, okra will not bear as prolifically as in warm climates. Red-podded cultivars, such as 'Burgundy', are often better producers.

DAYS TO MATURITY: 50–60 frost-free days.

HARVESTING AND STORING: Clip or pinch off young pods when they are 1–4 inches (2.5–10 cm) long and still soft. Larger pods will be woody. Harvest daily in warm weather. Freezes well.

SPECIAL TIPS: Okra blossoms also are edible.

CULTIVARS: 'Annie Oakley', 'Clemson Spineless', 'Emerald'.

Allium spp.   Liliaceae

ONION

A staple since before the pharaohs, the bulb of the versatile onion provides plenty of good eating in little garden space.

BEST CLIMATE AND SITE: All Zones for scallions or "green onions"; Zones 3 and warmer for bulb onions, choosing cultivars suited to day length. Full sun.

IDEAL SOIL CONDITIONS: Rich, well-drained, and humusy soil; pH 6.0–7.5.

GROWING GUIDELINES: For summer harvest, start indoors 3–4 months before last spring frost and set out 3–4 inches (8–10 cm) apart a month before last frost. Sow fall storage onions a month before last frost; thin to 3–4 inches (8–10 cm) apart. For quicker crops or in colder areas, grow onions from small bulbs called "sets," available at garden centers or by mail. Plant sets 1 inch (2.5 cm) deep and 2 inches (5 cm) apart, pulling every other one when ready for use as scallions. Grow onions in beds or small patches to maximize yield in small spaces. Keep well weeded; young onions have slender grass-like leaves and are easily shaded out. Irrigate and fertilize with fish emulsion or compost tea to encourage good early growth, which will determine eventual bulb size.

PEST AND DISEASE PREVENTION: Garden-grown onions have few pest or disease problems. Where onion maggots are troublesome, use row covers supported by hoops to prevent egg-laying by the onion maggot fly. Rotate plantings of onions and their relatives, such as garlic and leeks.

COMMON PROBLEMS: Cold weather may prevent bulb

Immature onions are often called scallions, but true scallions or bunching onions do not form bulbs.

formation, but failure to form bulbs more often indicates use of an unsuitable onion cultivar. Bulb formation is triggered by day length. Long-day cultivars are best in northern areas; short-day types in southern areas.

DAYS TO MATURITY: 60–115 days from transplanting. Scallions from sets may be harvested in as few as 35 days. Plants will withstand severe frost, but young ones may be set back.

HARVESTING AND STORING: Pull scallions and onions for fresh use as needed. For storage onions, wait until most tops have fallen over, then knock over any upright stalks with a rake. Pull the onions 1–2 days later and let them dry on the ground. (In wet weather, dry them on open mesh or shallow trays in a well-ventilated place.) Use onions with thick, green stems immediately; they will not store. When onions are thoroughly dry, braid the tops and hang in a cool, dry place or clip off the tops and store bulbs in mesh bags or slat-sided boxes.

SPECIAL TIPS: For space-saving scallions, sow several seeds in one pot and transplant later as a clump.

RELATED PLANTS:

EGYPTIAN ONION (*A. cepa,* Proliferum group): Also called tree onion, top onion, or catawissa onion. Edible, but strongly flavored. Instead of flowers, it produces a curious-looking head of small onion bulbs, which are planted in spring or fall to produce the next year's crop.

The tiny bulbs of some cultivars of onion are ideal for pickling or cooking whole as pearl onions.

Red onions are striking in salads. Like the yellow and white types, they can be grown from sets for quicker harvest.

PICKLING ONION (*A. cepa,* Cepa group): Also called pearl onion. Cultivars produce small bulbs, which are cooked or pickled whole, and include 'Snow Baby' and 'Early Aviv'.

POTATO ONION (*A. cepa,* Aggregatum group): Also called multiplier onion. Forms a cluster of underground bulbs. Usually harvested the second year after planting. Eat the larger bulbs and replant the smaller ones.

RED ONION (*A. cepa,* Cepa group): Also called Italian onion. These red-skinned cultivars are often sweeter than white- or yellow-skinned types and include 'Bennie's Red', 'Rossa di Milano', and 'Red Florence' (elongated shape).

SCALLION (*A. fistulosum*): Also called Welsh onion, ciboule, or Japanese bunching onion. Does not form a bulb, but rather a slender, white stalk. (Bulbing onion cultivars, pulled before the bulbs have begun to form, may also be used as scallions, or "green onions" as they are often called.) True scallion cultivars will self-divide, forming clumps that you can divide after their second year for a perennial scallion patch. They include 'Ishikura' and 'Evergreen Hardy White'.

SLICING ONION (*A. cepa,* Cepa group): Usually refers to very mild-flavored onions grown for fresh use. Most store poorly. Grow Vidalia types (like 'Texas Grano') in southern areas, and Walla Walla types (like 'Walla Walla Sweet') in northern ones.

SPANISH ONION (*A. cepa,* Cepa group): Large bulbs with mild flavor, usually yellow- or white-skinned. In favorable climates, bulbs can weigh up to 5 pounds (2.5 kg).

An interesting and decorative plant, the Egyptian onion produces clusters of bulbs instead of flowers.

Atriplex hortensis Chenopodiaceae

ORACH

The vivid foliage of red orach makes it an attractive addition to both ornamental and vegetable gardens.

BEST CLIMATE AND SITE: All Zones. Full sun.

IDEAL SOIL CONDITIONS: Not fussy, but prefers fertile, well-drained soil; pH 7.3–8.0.

GROWING GUIDELINES: Sow 4–6 weeks before last spring frost; thin to stand 8 inches (20 cm) apart, using thinnings in salad. Like spinach, orach bolts quickly in hot weather; unlike spinach, it will germinate and grow in warm weather. Keep seed heads pinched out for extended harvest. Frequent small plantings will assure plenty of young, tender leaves in spring, summer, and fall.

PEST AND DISEASE PREVENTION: Little troubled by pests.

COMMON PROBLEMS: Pull bolting plants, as they will stand 6 feet (1.8 m) or taller when in flower and may self-sow, becoming a weed problem.

DAYS TO MATURITY: 40–50 days; thinnings may be harvested sooner. Will withstand moderate frost.

HARVESTING AND STORING: Pick the succulent, young leaves as needed.

CULTIVARS: Named cultivars not generally available, but seed companies offer variants based on leaf color, usually dark green, light green, or red.

SPECIAL TIPS: Red orach (*A. hortensis* var. *atrosanguinea*) holds its color in cooking and is striking in salads and as garnish.

OTHER COMMON NAMES: French spinach, mountain spinach, sea purslane.

Petroselinum crispum var. *tuberosum* Umbelliferae

PARSLEY, HAMBURG

Grown for its thick, white taproot rather than its leaves, Hamburg parsley is attractive in the garden and tasty as well.

BEST CLIMATE AND SITE: Zones 4 and warmer. Full sun.

IDEAL SOIL CONDITIONS: Loose, deep, and well-drained soil, free of stones; pH 6.0–7.0.

GROWING GUIDELINES: Sow 4–6 weeks before last spring frost and keep moist until germinated, which can take several weeks. Thin to stand 6–8 inches (15–20 cm) apart. If started indoors, set plants out while still small to avoid curled or forked roots. Cultivate or mulch. Water regularly in dry spells to produce large, more tender roots.

PEST AND DISEASE PREVENTION: Handpick any celery- or parsleyworms, which are green caterpillars with black and yellow stripes.

COMMON PROBLEMS: Shallow or heavy soil can result in misshapen roots. Remove stones, break up clods, and don't step on seedbed. 'Early Sugar' is a shorter-rooted cultivar. Raised beds help.

DAYS TO MATURITY: 78–90 days. Will withstand severe frost.

HARVESTING AND STORING: Dig the root when it is large enough for your needs; frost sweetens the flavor. Keeps well in the ground (mulch to keep the soil diggable), or dig before hard freeze and store in damp sand or sawdust in a cool place.

SPECIAL TIPS: Dig overwintered plants before they begin to grow in spring. Roots shrink and toughen as the plants draw nutrients from them to set seed.

OTHER COMMON NAMES: Turnip-rooted parsley.

Pastinaca sativa   Umbelliferae

PARSNIP

Parsnips look rather like white carrots and have a distinctive taste. Cook them as you would carrots.

BEST CLIMATE AND SITE: All Zones. Full sun.

IDEAL SOIL CONDITIONS: Friable, open, well-drained soil; pH 6.0–7.0.

GROWING GUIDELINES: Sow after spring frosts have finished. Sow seeds directly where they are going to grow in a furrow 2½ inches (6 cm) deep and keep moist until germinated. Parsnip seeds take 21–28 days to germinate. Thin seedlings to 2–3 inches (5–8 cm) apart and control weeds by hand-weeding and cultivation. Do not overfeed with nitrogen fertilizers.

PEST AND DISEASE PREVENTION: Rarely troubled by pests.

COMMON PROBLEMS: Shallow, heavy soil or fresh manure in the soil may result in misshapen roots. Remove stones and break up clods before planting.

DAYS TO MATURITY: 126–140 days.

HARVESTING AND STORING: Start pulling roots early to spread the harvest. The remainder keep well in the soil. In winter cover the crown with a thick layer of straw mulch and mark the bed. Dig through mulch layer all winter for a continuous, fresh harvest. Roots store for a week or two on a shelf and for a few weeks in the refrigerator.

CULTIVARS: 'Hollow Crown', 'Harris Model'.

SPECIAL TIPS: If parsnips are grown on a well-prepared and fertilized bed, extra fertilizer is rarely needed, but liquid feeds, like compost tea, applied in mid-season, will promote faster growth.

Pisum sativum var. *sativum*   Leguminosae

PEA

Tall-growing peas will need some type of trellis to support their climbing stems.

BEST CLIMATE AND SITE: Zones 2 and warmer; grow as a spring and fall crop in most Zones, and as a winter vegetable in frost-free climates. Full sun on a site with good air circulation.

IDEAL SOIL CONDITIONS: Loose, well-drained soil; pH 6.0–7.0.

GROWING GUIDELINES: Peas do poorly in hot weather, so plant spring crops early to beat the heat unless your springs are long and cool. In Zones 7 and warmer, seed planted in late fall often overwinters and germinates in early spring. In colder Zones, plant 4–8 weeks before last spring frost. Sow fall crops 10–12 weeks before first fall frost. Plant bush types 1 inch (2.5 cm) deep and about 1 inch (2.5 cm) apart in double rows spaced 6 inches (15 cm) apart; plants will help support each other and yield will be greater in very little extra space. Plant tall types roughly 1 inch (2.5 cm) apart in single rows and provide a trellis for them to climb. Do not thin. Cultivate carefully to avoid uprooting plants, or mulch. Peas do not need supplementary fertilizer, but make sure you provide adequate moisture when they are in bloom.

PEST AND DISEASE PREVENTION: Rarely troubled by insects. Good air circulation and resistant cultivars help prevent powdery mildew. Do not plant where peas have grown the previous 2 years if root rot is a problem.

COMMON PROBLEMS: Fall plantings in humid areas,

The crisp pods of snap peas are also edible, providing a greater harvest from just a small planting.

Where space is limited, choose a pea like 'Tall Telephone'. You can grow them on teepees or trellises.

even of resistant cultivars, are often plagued by powdery mildew. Spray foliage thoroughly with a baking soda solution (1 teaspoon per quart [liter] of water).

DAYS TO MATURITY: 55–80 days. Young plants will withstand moderate frost, but fall crops will not mature after frost.

HARVESTING AND STORING: Pick shelling-type peas when pods are full and plump and peas are tender. Pick snap peas when pods are rounded but still smooth, before peas begin to bulge the sides of the pod. Pick snowpeas when pods are perfectly flat, showing only the tiniest hint of the pea inside. Peas are best eaten immediately or frozen or canned promptly, as their sugars begin to turn to starch quickly. Dry and store overmature crops as you would dry beans, for use in soups.

SPECIAL TIPS: Even low-growing cultivars benefit from some vine support. Stick twiggy brush cuttings into the pea row to serve as what old-time gardeners called "pea brush."

RELATED PLANTS:

SHELLING PEA (*P. sativum* var. *sativum*): Also called English pea, garden pea. Cultivars include 'Laxton's Progress' (bush type), 'Tall Telephone' (trellis type), and 'Wando' (bush type, heat-resistant).

SNAP PEA (*P. sativum* var. *sativum*): Pods are not fibrous like shelling peas and may be eaten cooked or raw in salads. They are best if the strings are removed before cooking. Cultivars include 'Sugar Snap' (trellis type), 'Sugar Ann' (bush type), and 'Sugar Daddy' (bush type, stringless).

SNOWPEA (*P. sativum* var. *macrocarpon*): Also called edible-podded or sugar pea. Favored in oriental cuisines. Cultivars include 'Oregon Sugar Pod' (trellis type), 'Mammoth Melting Sugar' (trellis type), and 'Little Sweetie' (bush type).

A staple in Asian cuisine, snowpeas should be picked when the peas inside are barely visible.

| *Arachis hypogaea* | Leguminosae | *Capsicum annuum* | Solanaceae |

PEANUT

PEPPER

The peanut is not a nut but a protein-rich vegetable in the legume family, usually roasted, boiled, or made into a sauce.

Bell peppers are crisp and juicy when green, but sweeter when allowed to ripen to red, yellow, or orange.

BEST CLIMATE AND SITE: Zones 7 and warmer, but may produce light crops in Zones 5–6. Full sun.

IDEAL SOIL CONDITIONS: Loose, well-drained, humusy soil with adequate calcium; pH 5.8–6.2.

GROWING GUIDELINES: In warmer climates, sow 1 inch (2.5 cm) deep and 3 inches (7.5 cm) apart after last spring frost when soil is warm, and thin to 12 inches (30 cm) apart. In cooler areas, start in individual pots 4–6 weeks before last frost and set out, 12 inches (30 cm) apart, in soil that has been prewarmed with black plastic mulch. Plant whole shells, or remove shell first, taking care not to damage the papery skin. Keep weeded. Do not mulch; if plastic mulch is used to warm soil, it must be removed when the peanuts flower. Stems bearing fertilized flowers dive into soil around the plant, and a peanut forms at the end of each stem.

PEST AND DISEASE PREVENTION: Little troubled.

DAYS TO MATURITY: 110–120 frost-free days.

HARVESTING AND STORING: Dig plants when frost has killed the foliage. Hang by the roots in a well-ventilated place until the pods have completely dried. Roast shelled nuts, or roast them in the shell by soaking clean pods in salted water for several hours, then heating in a 300°F (150°C) oven until completely dry and crisp (about 1 hour).

CULTIVARS: 'Jumbo Virginia', 'Spanish', 'Tennessee Red'.

OTHER COMMON NAMES: Goober, groundnut, pindar.

BEST CLIMATE AND SITE: Zones 4 and warmer. Full sun.

IDEAL SOIL CONDITIONS: Light, well-drained soil, not overly rich; pH 6.0–7.0.

GROWING GUIDELINES: Start indoors 6–8 weeks before last spring frost. Do not overwater pepper seedlings, as they are vulnerable to root rot. Set out 10–15 inches (25–38 cm) apart when frost danger is well past and the soil has warmed. In colder areas, prewarm the soil with black plastic mulch. Young peppers will tolerate cool spells but will not thrive until warmer weather arrives. Cultivate shallowly; do not mulch until the soil is thoroughly warm. Too much nitrogen will produce lush foliage and few peppers, but an application of fish emulsion or compost tea when the plants are in flower can help increase the yield. Magnesium is critical; in magnesium-poor soils, scatter 1 teaspoon of Epsom salts around the base of each plant. Irrigate in dry spells. Peppers, particularly thick-fleshed sweet peppers, are prone to blossom-end rot if drought-stressed.

PEST AND DISEASE PREVENTION: Do not plant where peppers or other nightshade family members, such as tomato and eggplant, have grown for 2 years. Protect young plants from cutworms with cardboard or foil collars. Keep peppers away from corn plantings to reduce earworm damage to peppers.

COMMON PROBLEMS: Sunscald can cause dry, sunken patches on the fruit. Plant more closely, keep

Colorful and sweet, bell peppers can add a festive touch to any summer meal.

The soul of many Asian and Mexican dishes, hot peppers come in many sizes, shapes, and degrees of "heat."

watered, and grow cultivars with good leaf canopies to avoid fruit damage.

DAYS TO MATURITY: 55–80 days from transplanting for green peppers, 15–20 days more for mature peppers.

HARVESTING AND STORING: Pick immature or green peppers when they are large enough for use. Leave some fruit on the plant to mature. Fully ripe peppers will be yellow, orange, or red, depending on the cultivar. Pick mature peppers when 50–75 percent colored; they will finish ripening in 1–2 days at room temperature. When frost threatens, harvest all remaining fruit. Fresh peppers will keep for 2 weeks or more if stored at around 55°F (13°C) (a refrigerator is too cold), or freeze them for winter cooking. Freeze or pickle thick-fleshed hot peppers such as jalapeño and hot cherry; dry thin-fleshed ones such as cayenne.

SPECIAL TIPS: Hot peppers and sweet peppers may cross-pollinate. Plant them well away from each other, especially if you intend to save seed. Peppers are picturesque plants of bushy habit that can be interplanted in the flower garden. They may also be grown in containers.

RELATED PLANTS:

HOT PEPPERS (*C. annuum,* Longum group): Also called chili peppers. Hot peppers come in a wide variety of shapes, sizes, and degrees of "heat." Milder cultivars include 'Ancho' or 'Poblano', a large, mildly hot pepper that is often served stuffed. Medium-hot peppers include jalapeño and Hungarian wax. Fiery peppers include cayenne and Thai pepper. Hot weather intensifies the flavor.

PICKLING PEPPERS (*C. annuum,* Conoides group): Usually small, thin-fleshed peppers with a cone shape, borne upright on the plant. 'Sweet Pickle' is ornamental, with red, orange, yellow, and purple fruits on the same plant. 'Pepperoncini' is an Italian favorite, long and thin, usually pickled when green but turns red at maturity.

SWEET PEPPERS (BELL) (*C. annuum,* Grossum group): Blocky in shape and thick-fleshed, most often used fresh in salads and relish trays or for stuffing. Most cultivars turn red or yellow when fully ripe; purple and "chocolate" peppers are colored at their immature stage and turn red when fully ripe. Cultivars include 'Ace' (red at maturity), 'Orobelle' (yellow), 'Jupiter' (red), and 'Oriole' (orange).

SWEET PEPPERS (OTHER) (*C. annuum,* Grossum group): Frying peppers, also called Italian, ramshorn, Cubanelle, or banana peppers, have elongated fruit up to 1 foot (30 cm) long. They are generally thinner-fleshed than bell peppers and hold their shape and flavor better in cooked dishes. Cultivars range from dark green to light yellow when immature and ripen to red, orange, or yellow; they include 'Biscayne' (pale green, ripens red), 'Sweet Banana' (yellow, ripens red), 'Bullhorn' (yellow,

PEPPER—CONTINUED

POTATO

Less juicy than bell peppers, frying peppers hold their shape and flavor better in cooked dishes.

Red potatoes are delicious simply boiled or baked in their jackets. They have moist, white flesh.

ripens orange), and 'Gypsy' (yellow, ripens red). Pimento peppers, also called cheese peppers, are thick-fleshed and flattened in shape. Excellent when fresh and the best pepper for roasting. Pimento cultivars include 'Super Red Pimento' and 'Yellow Cheese'.

The rich orange-yellow skin of 'Bullhorn' peppers contrasts perfectly with red tomatoes to make a colorful salad.

BEST CLIMATE AND SITE: Zones 3 and warmer. Full sun.

IDEAL SOIL CONDITIONS: Loose, well-drained, slightly acid soil with plenty of potash; ideal pH 5.2–5.7, but will also do well at pH 5.8–6.5.

GROWING GUIDELINES: Most potatoes do not come true from seed and are planted from pieces of tuber called "seed potatoes." Each piece should be about the size of an egg and contain one or more "eyes," or dormant buds. Cut seed potatoes to size and let the pieces dry for 1 day to avoid rotting in the ground. Plant 2–4 weeks before last spring frost, about a foot (30 cm) apart in a 3- to 4-inch (8–10 cm) deep trench. Rake soil level over seed potatoes. Tubers will form above the seed potato, not below it, so plants must be "hilled" to get good yields. With a hoe or rake, draw soil up around the plant when it is 6–8 inches (15–20 cm) tall, burying all but the topmost leaves. Repeat at least once more as the plant grows. Hilling also supports the plants and protects new potatoes from exposure to light. On level or slightly sloping beds, the furrow between rows can be used for irrigating. You can also grow potatoes under straw or leaf mulch, adding layers as the plant grows. In well-prepared and fertilized soil, no extra fertilizer is needed, but water regularly to promote smooth, well-developed potatoes.

PEST AND DISEASE PREVENTION: Rotate plantings

Fingerling potatoes are long and slender, with a waxy texture that makes them ideal for use in salads.

The leaves of your potato plants will tell you when to harvest: When the foliage dies down, dig tubers as needed.

of potato and its nightshade relatives, such as tomato and eggplant. Potatoes like soil that has recently been under sod, but prepare the area the fall before planting and till it once during the winter, if possible, to destroy grubs. Do not use fresh manure on potato beds; it encourages scab. Covering plants with row covers or dusting them with ground limestone helps deter flea beetles and Colorado potato beetles. A special strain of BT is available to control potato beetle larvae where the insect is a serious problem.

COMMON PROBLEMS: Exposed to light, potato tubers develop green patches that contain the toxic alkaloid solanine. Hill or mulch deeply to prevent sunlight from reaching the tubers; store harvested potatoes in a dark area.

DAYS TO MATURITY: 55–80 days. Young plants will withstand light frost.

HARVESTING AND STORING: Blooming is usually the signal that the plant has begun to form tubers. Check by gently probing the soil around the base of the plant, taking no more than a few tubers per plant for early use. "New potatoes" are those that have been freshly harvested and eaten within a day, before the skin toughens and sugars begin to convert to starch. When foliage dies back, potatoes are mature; dig as needed. Complete the harvest before hard freeze; potatoes that have been frozen will rot in storage. Dry without washing and store

in well-ventilated boxes or mesh bags in a cool to cold place, ideally 40°F (4°C).

SPECIAL TIPS: To speed up your harvest, try presprouting your seed potatoes. Set whole tubers in a cool, bright spot. Once they have developed short, bright sprouts (in about 4 to 6 weeks), plant the whole potatoes out into the garden as you would normally.

RELATED PLANTS:

FINGERLING POTATOES (*S. tuberosum*): Also called German potatoes. Long, slender tubers, often with a waxy texture that makes them ideal for potato salad. More prolific than standard potatoes. Dig soon after maturity; potatoes left in the ground tend to get "knobby." Cultivars include 'German Fingerling', 'Yellow Fingerling', and 'Larota'.

RED POTATOES (*S. tuberosum*): A red-skinned potato, usually with moist, white flesh, but some cultivars have cream-colored or yellow flesh. Small ones are often sold commercially as "new potatoes." Cultivars include 'Red Pontiac', 'Red Norland', 'Red Bliss', and 'Red Gold' (yellow flesh).

RUSSET POTATOES (*S. tuberosum*): Usually oblong, netted-skin potatoes with a mealy texture ideal for baking. Cultivars include 'Russet Burbank'.

WHITE POTATOES (*S. tuberosum*): White- or buff-colored skin and white flesh. All-purpose potato. 'Irish Cobbler' is a very thin-skinned and early cultivar. 'Kennebec' is adapted to many areas.

Cucurbita pepo var. *pepo*   Cucurbitaceae

PUMPKIN

White potatoes are good all-purpose potatoes. Like other types of potatoes, they can be stored for later use.

'Green Mountain' is an old and reliable cultivar. YELLOW POTATOES (*S. tuberosum*): Yellow-fleshed potatoes are gaining popularity for their beauty and delicate flavor. Cultivars include 'Bintje', 'Yukon Gold', and 'Yellow Finn'.

Potatoes thrive in loose, rich soil. To avoid potato scab, don't plant in beds that contain fresh manure.

As familiar a sight in fall as falling leaves, pumpkin makes fine pies and baked dishes, as well as jack-o'-lanterns.

BEST CLIMATE AND SITE: Zones 4 and warmer. Full sun.

IDEAL SOIL CONDITIONS: Not fussy about texture, but likes a fertile soil; pH 6.0–7.0.

GROWING GUIDELINES: Sow after last spring frost, two or three seeds to a hill, in hills 4–5 feet (1.2–1.5 m) apart. In short-season areas, start pumpkins indoors in individual pots 2–3 weeks before last frost. Set out carefully to avoid disturbing the roots. A good shovelful of compost or well-rotted manure in each hill will boost growth. Mulch or cultivate until the plants begin to vine. Pumpkin's broad leaves will shade out most weeds.

PEST AND DISEASE PREVENTION: Row covers deter cucumber beetles and squash borers, but remove them when plants bloom to allow pollination. Straw mulch also helps deter cucumber beetles. Prevent mildew by providing good air circulation.

COMMON PROBLEMS: Plants that collapse before setting fruit are probably victims of the squash borer. Those that collapse before fruit ripens may be infected with a wilt disease spread by cucumber beetles. Cover plants until they bloom to reduce risk of crop loss.

DAYS TO MATURITY: 100–115 frost-free days. For homegrown jack-o'-lanterns in short-season areas, grow a cultivar such as 'Autumn Gold', which is orange when immature.

HARVESTING AND STORING: Harvest pumpkins

Japanese pumpkin, a winter squash, is green with orange flesh. It is often used in oriental cooking.

when they are fully colored and their shell is hard, or after light frost has killed the vines. Store in a dry, cool area. Pumpkins keep for several months but lose flavor in long storage. The cooked flesh freezes well, however. Cultivars meant for cooking, such as 'Small Sugar', are better eating than large field pumpkins.

RELATED PLANTS:

JAPANESE PUMPKIN (*C. mixta*): Also called Hokkaido squash or Chinese pumpkin. Green-skinned with dark orange, moist, sweet flesh. Not for carving, but a superb winter squash.

Portulaca oleracea var. *sativa*   Portulacaceae

PURSLANE

Popular in Europe as a salad green, cultivated purslane is taller and more succulent than its cousin the weed.

BEST CLIMATE AND SITE: Zones 4 and warmer. Full sun.

IDEAL SOIL CONDITIONS: Not fussy; will even grow in sand and is drought-resistant; pH 6.0–7.0.

GROWING GUIDELINES: Seed thinly in rows or broadcast in small beds after frost danger is past in spring. Do not cover seeds but firm gently into the soil with the back of a hoe. Keep moist until seeds germinate. Thin to 4–6 inches (10–15 cm) apart; use thinnings in salads. Sow small successive crops up to 2 months before fall frost. Water frequently.

PEST AND DISEASE PREVENTION: Little bothered by pests.

DAYS TO MATURITY: 60–70 days, although thinnings may be eaten sooner.

HARVESTING AND STORING: Harvest fresh leaves and stems with scissors as needed, leaving 1 inch (2.5 cm) or more aboveground to sprout new leaves. May be harvested four or five times. Best used fresh; does not store well.

SPECIAL TIPS: Common purslane is also edible. Collect seeds of the best wild plants to grow in the garden, saving seed each year from plants with the best flavor and growth habit.

CULTIVARS: Named cultivars are not generally available, but seed suppliers may offer variants based on leaf size or color, usually green or golden.

OTHER COMMON NAMES: Pourpier, pusley, verdolaga.

RADISH

Fast-growing radishes add a crisp, colorful touch to spring salads. Harvest them while they are young and tender.

BEST CLIMATE AND SITE: All Zones; grow as a winter vegetable in mild areas.

IDEAL SOIL CONDITIONS: Loose, moisture-retentive soil; pH 5.5–6.8.

GROWING GUIDELINES: Sow spring radishes 3–5 weeks before last spring frost, ½ inch (12 mm) deep in double or triple rows. Make the rows short and sow small successive crops every 2 weeks until a month after frost. Sow fall crops starting 8 weeks before fall frost and continuing until frost. Thin spring radishes to 1–2 inches (2.5–5 cm) apart, or seed sparingly and thin by pulling radishes as they reach eating size. Sow winter radishes 8–10 weeks before fall frost and thin to 4–6 inches (10–15 cm) apart. Keep radishes moist to avoid strong flavor and toughness. Mulch between rows to keep soil moist in hot weather.

PEST AND DISEASE PREVENTION: Rotate radishes and other root crops to reduce damage from root maggots. Spring crops are more vulnerable than fall crops.

COMMON PROBLEMS: Spring radishes, especially slender French breakfast types, become woody and bitter when overmature. Plant small but frequent crops to keep tender radishes on the table. The roots of radishes planted in heavy soils are apt to be misshapen and have a number of branching side roots.

DAYS TO MATURITY: 21–35 days for spring radishes;

Black Spanish radishes are very pungent and keep longer than other radishes. The flesh is white.

50–60 days for winter radishes. Will withstand moderate frost.

HARVESTING AND STORING: Pull spring and fall crops when large enough for use. Pull winter radishes as needed when they reach eating size; harvest all roots before hard freeze. Store in damp sand or sawdust in a cool place, or pickle.

SPECIAL TIPS: Interplant spring radishes with slower-growing crops such as broccoli and cabbage. You will harvest the radishes before the companion crop needs the space. Because of their quick germination and rapid growth, radishes are frequently mixed with beet or carrot seeds: The radishes germinate first and therefore mark the rows of the slower-growing crops. Because they grow quickly, they are a good crop for children to grow.

RELATED PLANTS:

BLACK SPANISH RADISH (*R. sativus* var. *longipinnatus*): Up to 4 inches (10 cm) in diameter, with black skin and white flesh. Very pungent; a good keeper.

CHINESE RADISH (*R. sativus* var. *longipinnatus*): Also called lo bok or lo po. Long and white like the Japanese radish, but sweet-flavored, not pungent. Oriental seed suppliers offer named cultivars or variants based on skin color.

JAPANESE RADISH (*R. sativus* var. *longipinnatus*): Also called daikon or daiko. White, carrot-shaped radish, up to 2 feet (60 cm) long. Japanese radishes

Rheum rhabarbarum   Polygonaceae

RHUBARB

The Japanese radish, or daikon, is pungent and is often grated for use as a condiment.

Rhubarb's stout leaf stalks, by themselves or with seasonal fruits, make delicious pies, cobblers, sauces, and preserves.

are often grated for use as a condiment, flavored with soy and ginger. Cultivars include 'Miyashige' and 'Summer Cross'.

WINTER RADISH (*R. sativus* var. *longipinnatus*): A catch-all term for radish cultivars that are suitable for at least short-term storage as a root vegetable. In addition to those above, the German beer radish (large, white, and turnip-shaped) is popular in Europe.

BEST CLIMATE AND SITE: Zones 8 and cooler. Full sun.

IDEAL SOIL CONDITIONS: Rich, deep, well-drained soil; pH 5.0–6.8.

GROWING GUIDELINES: Can grow from seed, but it is quicker to use divisions of this hardy perennial. Plant in early spring, 1–2 months before last spring frost. Dig a hole big enough to accommodate roots and cover the crown with no more than 2 inches (5 cm) of soil. Mix a shovelful of compost or well-rotted manure into the soil to get the plant off to a good start. Cultivate or mulch. Heavy feeder; top-dress every spring with compost or rotted manure. Drought-tolerant, but provide adequate moisture to ensure abundant, tender stalks.

PEST AND DISEASE PREVENTION: Little troubled by pests. Plant in well-drained soil to avoid root rot.

COMMON PROBLEMS: Cut off flower stems that rob plant of energy, resulting in stringy leaf stalks.

DAYS TO MATURITY: Harvest sparingly the second year after planting and more heavily thereafter.

HARVESTING AND STORING: To harvest large, outer leaf stalks, grasp them near the base and pull with a slight twisting motion. *Remove the leaves, which are toxic.* Harvest sparingly in fall as well as in spring, but spring stalks are more tender. Chop and freeze when plentiful.

CULTIVARS: 'MacDonald', 'Valentine'. 'Victoria' is robust, and green-stemmed rather than red.

RUTABAGA

A favorite in Scandinavian countries, rutabaga is a rugged garden vegetable that shrugs off cold and adverse conditions.

BEST CLIMATE AND SITE: Zones 3 and warmer; grow as a winter vegetable in mild areas. Full sun.

IDEAL SOIL CONDITIONS: Fertile, well-drained soil. Will tolerate heavy soil better than most other root vegetables; pH 5.5–6.8.

GROWING GUIDELINES: Sow 12–14 weeks before first fall frost and thin to stand 4–6 inches (10–15 cm) apart. Cultivate early; the plant's large leaves will quickly grow to shade out weeds.

PEST AND DISEASE PREVENTION: Rotate rutabaga and other root crops to avoid root maggots. Protect young plants from flea beetles with row covers; larger plants may suffer cosmetic damage that will not affect yield.

COMMON PROBLEMS: Rutabagas that mature in hot weather may be tough.

DAYS TO MATURITY: 90–110 days. Will withstand severe frost.

HARVESTING AND STORING: Pull rutabagas when they are large enough for use. The greens also are edible. Harvest all roots before hard freeze and store in damp sand or sawdust in a cool place. Rutabagas may be waxed to prevent wrinkling during storage. Trim the root and crown and dip them briefly in a pot of water with a layer of melted paraffin on top.

CULTIVARS: 'Altasweet', 'Laurentian'.

OTHER COMMON NAMES: Swede, Swede turnip, yellow turnip.

SALSIFY

A hardy root vegetable, salsify looks like a parsnip with narrow, grassy leaves, but has a delicate, oyster-like flavor.

BEST CLIMATE AND SITE: Zones 4 and warmer; grow as a winter vegetable in mild climates. Full sun.

IDEAL SOIL CONDITIONS: Deep, light, and rich soil; pH 6.0–8.0.

GROWING GUIDELINES: Sow ½ inch (12 mm) deep 2–4 weeks before last spring frost, or in fall in mild climates. Thin to stand 3–4 inches (8–10 cm) apart. Water regularly; mulch helps maintain soil moisture and produce smoother, more tender roots.

PEST AND DISEASE PREVENTION: Little troubled by pests or diseases.

COMMON PROBLEMS: Spotty or slow germination may indicate old seed. Purchase fresh salsify seed each year.

DAYS TO MATURITY: 120–150 days. Young plants will withstand light frost; mature ones, severe frost.

HARVESTING AND STORING: Dig roots as needed when large enough. Roots are not damaged by freezing, but mulch the bed before hard freeze to keep the ground diggable. Store harvested roots in damp sand or sawdust in a cool place.

RELATED PLANTS:

BLACK SALSIFY (*Scorzonera hispanica*): Cream root with black skin. Leaves may be eaten in salads.

SPANISH SALSIFY (*Scolymus hispanicus*): Also called golden thistle. Milder flavor than common salsify.

OTHER COMMON NAMES: Oyster plant, vegetable oyster.

Allium cepa, Aggregatum group | Liliaceae

SHALLOT

An onion relative, shallots produce small, firm bulbs that keep like garlic but are much milder in flavor.

BEST CLIMATE AND SITE: All Zones; may be fall-planted in Zones 6 and warmer. Full sun.

IDEAL SOIL CONDITIONS: Rich, humusy, well-drained soil. Will tolerate all but the most acid soil.

GROWING GUIDELINES: Shallots do not grow from seed but from bulblets or "sets." Plant 2–4 weeks before last spring frost, 1 inch (2.5 cm) deep and 4–6 inches (10–15 cm) apart. Keep cultivated or mulch and water regularly to encourage strong early growth. Each set will divide and produce 8–10 shallots. Where climate permits, fall planting will produce larger shallots the following summer.

PEST AND DISEASE PREVENTION: To avoid root maggots, do not plant where shallots or their relatives, such as onions or leeks, have grown the previous year.

COMMON PROBLEMS: Dry conditions or poor soil produces scrawny shallots. Work in plenty of compost or well-rotted manure and water regularly.

DAYS TO MATURITY: 120–150 days; or pull earlier to eat as scallions. Will withstand moderate frost.

HARVESTING AND STORING: When the tops are nearly dry, pull plants and dry the bulbs in a well-ventilated, sunny area. Store by hanging in a cool, dry place, or clip the stems and store the bulbs in mesh bags.

CULTIVARS: Named cultivars not generally available; most seed suppliers offer generic "French shallots."

Sium sisarum | Umbelliferae

SKIRRET

Skirret is an unusual perennial that forms clusters of swollen, white roots, eaten after the woody central core is removed.

BEST CLIMATE AND SITE: All Zones. Full sun.

IDEAL SOIL CONDITIONS: Deep, rich, and moisture-retentive soil; pH 6.0–6.7.

GROWING GUIDELINES: Sow seed 4–6 weeks before last spring frost, or in fall for harvest the following year—or start indoors. Thin to stand 12 inches (30 cm) apart. Water generously. Save a few rootlets from the plants you harvest, and replant them for next year's crop.

PEST AND DISEASE PREVENTION: Little troubled by pests.

COMMON PROBLEMS: Some plants will produce very woody roots. Dig out and discard those, selecting rootlets of less-woody plants to plant for next year's crop.

DAYS TO MATURITY: 120–150 days for first-year harvest; thereafter you can harvest roots each year from fall through early spring.

HARVESTING AND STORING: Dig roots as needed, from fall until new growth begins in the spring. Mulch the bed to keep the ground diggable. Skirret stores best in the ground.

CULTIVARS: Named cultivars not generally available.

SORREL, FRENCH

A tender green with a lemony flavor, delicious in salads, French sorrel withstands dry weather better than other sorrels.

BEST CLIMATE AND SITE: All Zones. Full sun or partial shade.

IDEAL SOIL CONDITIONS: Rich, moist soil; pH 6.0–7.0.

GROWING GUIDELINES: Start indoors or sow directly in the garden 2–4 weeks before last spring frost. Thin to stand 6–8 inches (15–20 cm) apart, using thinnings in salads. Cultivate or mulch and keep watered for best production. Top-dress the bed with compost or well-rotted manure each fall or spring. Sorrel is perennial but will decline after 3–4 years. Start a new bed from seed or divide existing plants.

PEST AND DISEASE PREVENTION: Use row covers to protect plants from leaf-eating insects.

COMMON PROBLEMS: Keep flower stems cut back to extend spring harvest.

DAYS TO MATURITY: 60 days for first-year crop, although thinnings may be harvested sooner. Will withstand moderate frost.

HARVESTING AND STORING: Pick fresh leaves as needed. Produces abundantly in spring and fall; less productive in hot weather. Freeze leaves for use in soup and sauces.

RELATED PLANTS:

GARDEN SORREL (*R. acetosa*): Also called sour dock. Longer leaves and milder flavor than French sorrel.
SHEEP'S SORREL (*R. acetosella*): Also called red sorrel. Most common "weed" sorrel. Edible but invasive.

SOYBEAN

One of the oldest food crops, soybean is widely used in oriental cuisines, either fresh or prepared as tofu, soy sauce, or miso.

BEST CLIMATE AND SITE: Zones 5 and warmer. Full sun.

IDEAL SOIL CONDITIONS: Light, well-drained soil; pH 6.0–7.0.

GROWING GUIDELINES: Sow after frost danger is past, 1 inch (2.5 cm) deep and 2 inches (5 cm) apart. Keep cultivated until plants grow large enough to shade out weeds. Water is critical when plants are in flower.

PEST AND DISEASE PREVENTION: Less troubled by pests than other beans. Rotate plantings of soybean and its legume relatives, such as beans and peas.

COMMON PROBLEMS: May not mature for use as a dry bean in short-season areas, but some cultivars can be grown for fresh use.

DAYS TO MATURITY: 65–120 frost-free days.

HARVESTING AND STORING: Harvest for fresh use when beans are plump and tender. Can or freeze fresh beans. For dry soybeans, pull the plants when most of the foliage has died and hang them to finish drying in a well-ventilated place. Shell and store as you would any dry bean.

CULTIVARS: 'Envy' and 'Hakucho Early' are short-season cultivars, primarily for fresh use. 'Panther' and 'Black Jet' need a longer season and are grown primarily for use as dry beans. Seed companies sometimes offer variants by seed color, usually yellow, green, or black.

Spinacia oleracea   Chenopodiaceae

SPINACH

Delicious and healthy fresh or cooked, spinach needs cool weather and plenty of water to make abundant crisp leaves.

BEST CLIMATE AND SITE: All Zones. Full sun or partial shade.

IDEAL SOIL CONDITIONS: Moist, fertile, well-limed soil; pH 6.0–7.0.

GROWING GUIDELINES: Sow ½ inch (12 mm) deep, 4–6 weeks before last spring frost. In very short-season areas, or where hot weather sets in abruptly, start spinach indoors. Set out or thin to 4–6 inches (10–15 cm) apart. Keep weed-free and water regularly. Sow fall crops 4–6 weeks before first frost, or later for overwintering. Seed fall crops heavily, as spinach germinates poorly in warm soil.

PEST AND DISEASE PREVENTION: Use row covers to protect from leaf miners and chewing insects.

COMMON PROBLEMS: Hot weather and lengthening days can cause plants to bolt (go to seed). Use heat-resistant cultivars and sow spring crops early to avoid this problem.

DAYS TO MATURITY: 40–53 days, although thinnings may be harvested sooner. Spinach will withstand moderate frost.

HARVESTING AND STORING: Pick larger outside leaves or harvest whole plant at its base. Spinach freezes well for use as a cooked vegetable.

CULTIVARS: 'Melody', 'Longstanding Bloomsdale', 'Tyee'.

Basella alba   Basellaceae

SPINACH, MALABAR

A vining plant native to tropical Africa and Asia, Malabar spinach has glossy leaves that are used like spinach.

BEST CLIMATE AND SITE: Zones 7 and warmer. Full sun, near a fence or trellis.

IDEAL SOIL CONDITIONS: Moist, rich, humusy soil; pH 6.0–7.5.

GROWING GUIDELINES: Start indoors up to 8 weeks before the last frost and set out after frost danger is well past. Likes heat and grows slowly if at all until the temperature is to its liking. Set plants 3 feet (90 cm) apart and provide a fence or trellis. Water well in dry weather.

PEST AND DISEASE PREVENTION: Little troubled by pests.

COMMON PROBLEMS: In warm climates, can grow to 30 feet (9 m) in a season, overrunning small gardens, but 6–10 feet (1.8–3 m) is more common.

DAYS TO MATURITY: 120–150 frost-free days, but some leaves may be harvested earlier.

HARVESTING AND STORING: Pick fresh leaves as needed; in ideal weather the vine regrows rapidly.

CULTIVARS: 'Rubra' has red-veined leaves and red stems.

OTHER COMMON NAMES: Ceylon spinach, Indian spinach.

Tetragonia tetragonioides   Tetragoniaceae

SPINACH, NEW ZEALAND

New Zealand spinach is a branching, mat-like plant that thrives in weather too hot for true spinach.

BEST CLIMATE AND SITE: Zones 4 and warmer. Full sun.

IDEAL SOIL CONDITIONS: Rich, well-drained soil; pH 6.5–7.5.

GROWING GUIDELINES: Soak seed overnight to hasten germination and sow directly in the garden 1–2 weeks before last spring frost. In warm climates, prechill seed in the refrigerator for 1–2 days. In short-season areas, start indoors 4–6 weeks before the last frost. Set plants 12–18 inches (30–45 cm) apart when frost danger is past.

PEST AND DISEASE PREVENTION: Rarely troubled by pests.

COMMON PROBLEMS: Seed can be slow to germinate. Mark the row well to avoid cultivating out young plants.

DAYS TO MATURITY: 55–70 days.

HARVESTING AND STORING: Pick about 4–6 inches (10–15 cm) of branch tips, together with the leaves, which are small and brittle. Whole plants may be cut above the ground when they are small; the stem will resprout. Use as you would spinach.

SPECIAL TIPS: New Zealand spinach is good for hot, dry climates where true spinach does poorly. Tolerates saline soils.

CULTIVARS: Named cultivars not generally available.

Ipomoea reptans   Convolvulaceae

SPINACH, WATER

A member of the morning glory family, water spinach is a perennial in its native subtropics. Eat young leaves and shoots.

BEST CLIMATE AND SITE: Zones 7 and warmer. Full sun or partial shade.

IDEAL SOIL CONDITIONS: Rich, constantly moist soil; pH 5.5–6.5.

GROWING GUIDELINES: Start indoors and set out after frost danger is well past. Allow plenty of space; sends out runners and roots at the leaf nodes, quickly becoming a clump. Keep the soil quite moist, as water spinach is a semi-aquatic plant.

PEST AND DISEASE PREVENTION: Not generally troubled by pests.

COMMON PROBLEMS: Water spinach needs day temperatures over 72°F (22°C) to grow quickly. Do not plant too early.

DAYS TO MATURITY: 100–120 days. Poor growth in cool seasons; will not withstand frost.

HARVESTING AND STORING: Pick young leaves and shoots for use in salads, stir fries, or cooked dishes.

CULTIVARS: Named cultivars not generally available. Oriental seed suppliers may offer variants based on growth habit or leaf size.

OTHER COMMON NAMES: Swamp cabbage, kankon.

SPROUTS

A *wide variety of seeds, including alfalfa, are easy to sprout in water and are tasty cooked or eaten raw in salads.*

BEST CLIMATE AND SITE: All Zones. Sprouts are grown indoors.

IDEAL SOIL CONDITIONS: No soil is required for sprouted seeds.

GROWING GUIDELINES: Use only seeds not treated with fungicides or other chemicals; natural food stores often carry seeds packaged especially for sprouting. Use fresh seeds to ensure good germination. Soak the seeds overnight in water. Pick out broken seeds and rinse the rest several times in warm water in a strainer or colander. Put the seeds in a wide-mouthed jar, cover it with cheesecloth, and invert it over a dish to catch any remaining rinse water. Keep in a warm place, rinsing the seeds several times daily with warm water.

COMMON PROBLEMS: Old seeds may germinate slowly or not at all.

HARVESTING AND STORING: Sprouts are ready in 3–5 days, or slightly longer if you are sprouting hard-seeded vegetables. Store sprouted seeds in the refrigerator and use promptly; many sprouts mold quickly.

CULTIVARS: Many vegetable and grain seeds are excellent for sprouting, including wheat, barley, rye, oats, peas, millet, sunflowers, beans, radishes, and broccoli. Flavor and texture vary widely; experiment to find out what you like best.

Cucurbita pepo var. *melopepo*   Cucurbitaceae

SQUASH, SUMMER

Pattypan squash is disk-shaped with scalloped edges, and is especially attractive for serving as a stuffed vegetable.

BEST CLIMATE AND SITE: Zones 3 and warmer. Full sun in a site with good air circulation.

IDEAL SOIL CONDITIONS: Rich, well-drained, humusy soil; pH 6.0–6.5.

GROWING GUIDELINES: Sow when all danger of frost is past, or start indoors in individual pots 2–3 weeks before last frost and transplant carefully to avoid breaking roots. Space bush types 18–24 inches (45–60 cm) apart; vining types need wider spacing, up to 4 feet (1.2 m), and should be planted in hills of three or four seeds. Cultivate or mulch. Squash needs ample moisture to keep the fruit coming, and it benefits from plenty of compost or rotted manure.

PEST AND DISEASE PREVENTION: Use row covers to protect young squash plants from cucumber beetles and squash borers; remove covers when the plants bloom. Radishes or basil interplanted with squash helps repel beetles and squash bugs. Do not plant where squash or its relatives, like melon and cucumber, have grown the previous year. Provide good air circulation to avoid mildew. If plants are attacked by mildew, spray foliage thoroughly with a mild baking soda solution (1 teaspoon per quart [liter] of water). You can reduce the chance of many squash diseases by planting resistant cultivars. A straw mulch will help to deter insects and keep the soil moist, which is essential for healthy fruit. Dusting the foliage

Zucchini is best picked when less than 8 inches (20 cm) long, but larger fruits are delicious in soups, casseroles, and breads.

Colorful 'Golden Crookneck' squash looks as interesting on the table as in the garden. It thrives in warm weather.

with wood ashes can also deter insect pests.

COMMON PROBLEMS: Fruit that turns black and rots before reaching picking size has not been pollinated. This often happens early, before male blossoms appear, or in cool spells, when pollinating insects are less active.

DAYS TO MATURITY: 42–65 frost-free days.

HARVESTING AND STORING: Pick summer squash small for best flavor: 4–8 inches (10–20 cm) for zucchini, 4–5 inches (10–13 cm) for yellow squash, teacup-sized for pattypans. Harvest frequently to encourage more fruiting. Cut squash from the plant with a sharp knife and handle carefully, as the skin bruises easily. Freeze or pickle excess fruit. Blossoms are also edible, and are good stuffed and baked or fried. Pick blossoms in early morning, before they have been pollinated.

SPECIAL TIPS: Successive small plantings of just a few plants, sown a month apart, will keep the harvest manageable and extend the picking season where squash often succumbs to insects and disease. Vining types of summer squash may be encouraged to grow up a trellis to save space. In small gardens grow vining squash on a fence at the back of the ornamental garden.

RELATED PLANTS:

PATTYPAN SQUASH (*C. pepo* var. *melopepo*): 'Sunburst', 'Early White Bush', 'Peter Pan Bush Scallop'.

YELLOW SQUASH (*C. pepo* var. *melopepo*): 'Seneca Prolific' (straightneck), 'Pic-n-Pic Hybrid' (crookneck), 'Butterstick' (straightneck), and 'Golden Crookneck'.

ZUCCHINI (*C. pepo* var. *melopepo*): 'Aristocrat' (dark green), 'Gold Rush' (yellow), 'Cocozelle' (Italian type, green with light stripes).

Cucurbita spp.   Cucurbitaceae

SQUASH, WINTER

Butternut squash is light tan, with sweet, orange flesh. It ripens more quickly than many other winter squashes.

The flesh of 'Gold Nugget' squash is rich in flavor. It is faster growing than other squashes.

BEST CLIMATE AND SITE: Zones 4 and warmer. Full sun, in an area with good air circulation.

IDEAL SOIL CONDITIONS: Rich, well-composted, well-drained soil; pH 6.0–6.5.

GROWING GUIDELINES: Sow 1–2 weeks after the last spring frost, 18–24 inches (45–60 cm) apart in rows or in hills 4–5 feet (1.2–1.5 m) apart with three or four plants in a hill. In short-season areas, start indoors in individual pots 2–3 weeks before planting and set into soil that has been prewarmed with black plastic mulch. Transplant carefully to avoid damaging the roots. Mulch to hold in soil moisture.

PEST AND DISEASE PREVENTION: Use row covers to protect young plants from cucumber beetles and squash borers, but remove when plants are in bloom. Straw mulch will help deter insects, as will interplantings of radishes or basil. Provide good air circulation to avoid mildew. Do not plant where squash or its relatives, like melon and cucumbers, have grown the previous year.

COMMON PROBLEMS: Mildew can be a serious problem, especially in late-summer damp spells. Plants may be killed, or weakened too much to ripen fruit. At first sign of mildew, spray foliage thoroughly with a mild baking soda solution (1 teaspoon per quart [liter] of water).

DAYS TO MATURITY: 85–110 frost-free days.

HARVESTING AND STORING: Harvest when the shell is hard enough that it cannot be dented with a fingernail (except vegetable marrow squash. Harvest when 8–12 inches [20–30 cm] long and prepare like summer squash). Harvest all fruit as soon as light frost has killed the vines; more than slight frost will shorten storage life. Leave on the ground to "cure" in the sun for 10–14 days; cover the fruit if frost threatens. This curing sweetens the flesh and toughens the skin for storage. Wipe the skin with a cloth dipped in a weak bleach solution to help prevent rot. Store in a cool, well-ventilated place.

RELATED PLANTS:

ACORN SQUASH: (*C. pepo* var. *pepo*): Also called pepper squash. Acorn-shaped, usually dark green to near black. Sweet, moist, orange flesh. High-yielding, usually five to seven fruits per plant. 'Ebony Acorn', 'Table Queen', 'Swan White', and 'Cream of the Crop' (white-skinned with off-white flesh).

BUTTERCUP SQUASH (*C. maxima*): Dark green and blocky, with a "button" or small turban at blossom end. Orange-fleshed fruits weigh 3–5 pounds (1.5–2.5 kg) and keep well. The Japanese kabocha squashes are similar. Cultivars include 'Sweet Mama' and 'Honey Delight'.

BUTTERNUT SQUASH (*C. moschata*): Small seed cavity with a thick neck of solid, orange flesh. An excellent keeper, 4–5 pounds (2–2.5 kg). Cultivars include 'Waltham Butternut' and 'Ponca'.

Teacup-sized and just right for single servings, 'Sweet Dumpling' squash are ivory with dark green stripes.

Serve the spaghetti-like strands of 'Vegetable Spaghetti' with a cheese sauce, or your favorite pasta sauce.

DELICATA SQUASH (*C. pepo*): Also called sweet potato squash. Oblong, 2-pound (1 kg) fruits striped green and ivory, 8 inches (20 cm) long and 3 inches (7.5 cm) in diameter. Orange flesh. Doesn't need curing. Cultivars include 'Delicata' and 'Sugarloaf'.

'GOLD NUGGET' SQUASH (*C. maxima*): This quick-maturing winter squash has round orange fruits weighing up to 2 pounds (1 kg). A bush-type plant good for small gardens.

HUBBARD SQUASH (*C. maxima*): Also called blue hubbard. Large, warty fruits, usually one or two to a vine, with hard rind. An excellent keeper. Nonstringy flesh may be frozen after being cooked and puréed. Cultivars include 'New England Blue Hubbard' and 'Hubbard Improved Green'.

SPAGHETTI SQUASH (*C. pepo*): Also called vegetable spaghetti. Oval-shaped, with buff or light yellow skin and pale orange-yellow flesh. Bake, then scrape out the flesh with a fork. It separates into pasta-like strands; serve topped with tomato sauce. Cultivars include 'Vegetable Spaghetti' and 'Orangetti Hybrid'.

'SWEET DUMPLING' SQUASH (*C. pepo*): Flattened, round fruit, 3–4 inches (8–10 cm) in diameter, with light orange flesh and attractive striped skin. Requires no curing before storage.

VEGETABLE MARROW SQUASH (*C. pepo* var. *pepo*): Also called Lebanese zucchini. There are bush forms, but most cultivars are trailing vines that can be trellised. Pick at 8–12 inches (20–30 cm), and use the dense, meaty fruits like summer squash or eggplant. Cultivars include 'Cousa' and 'Vegetable Marrow Bush'.

Unlike most cultivars of acorn squash, 'Swan White' has creamy-white skin and pale yellow flesh.

| *Helianthus annuus* | Compositae | *Ipomoea batatas* | Convolvulaceae |

SUNFLOWER

SWEET POTATO

Sunflowers grow up to 10 feet (3 m) tall with showy, yellow-petaled flowers. They produce nutty-flavored seeds.

Sweet potatoes are a warm-weather crop that produces tasty and nutritious tuberous roots.

BEST CLIMATE AND SITE: Zones 5 and warmer. Full sun.

IDEAL SOIL CONDITIONS: Not fussy, but likes rich soil; pH 6.0–7.5.

GROWING GUIDELINES: Sow after frost danger is past, ½ inch (12 mm) deep and 6 inches (15 cm) apart. Thin to stand 18–24 inches (45–60 cm) apart. In short-season areas, start indoors 2–3 weeks before the last frost, and set plants out after frost danger is past. Cultivate or mulch. Drought-tolerant, but regular watering will produce larger seed heads.

PEST AND DISEASE PREVENTION: Provide good air circulation to avoid mildew. As the heads mature, you may have to cover them with cheesecloth to deter marauding birds and animals.

COMMON PROBLEMS: Sunflowers bloom relatively quickly but take a long time to ripen their seeds. In colder areas, choose an early-maturing cultivar.

DAYS TO MATURITY: 95–165 frost-free days.

HARVESTING AND STORING: When the back of the seed head is dry or frost has killed the plant, harvest the entire head with about 1 foot (30 cm) of stalk and hang or lay it on newspaper to dry in a well-ventilated area. Rub the seed head to dislodge the seeds and store them in air-tight jars in a cool place.

SPECIAL TIPS: Very heavy heads may need support.

CULTIVARS: 'Jumbo', 'Mammoth Russian', 'Sundak'.

BEST CLIMATE AND SITE: Zones 5 and warmer. Full sun.

IDEAL SOIL CONDITIONS: Loose, well-drained soil that is not too rich; pH 5.5–6.5.

GROWING GUIDELINES: Plant from growing shoots or "slips," which are available by mail or at many garden centers. Plant after frost danger is well past. With a rake, make a ridge of soil 6–10 inches (15–25 cm) high and 6–8 inches (15–20 cm) wide. Plant slips into the ridge about 12 inches (30 cm) apart. Mound soil onto the ridge at least once before the vining plants make further cultivation impossible. Keep newly set slips watered until they begin to grow. After that, irrigate sweet potatoes only in extended dry spells.

PEST AND DISEASE PREVENTION: Use disease-free slips. Rotate plantings and keep soil organic matter high to reduce nematode damage.

COMMON PROBLEMS: Deer love the tender leaves and shoots; if they are a problem, you may need to fence off your sweet potatoes.

DAYS TO MATURITY: 90–120 frost-free days.

HARVESTING AND STORING: Dig sweet potatoes before frost or as soon as the vines have been killed by a light frost. Left in the ground, they will spoil. The tuberous roots are very tender, so dig and handle them carefully. Cure them for 10–14 days in a warm, fairly humid area, and store where the temperature does not fall below 55°F (13°C).

SWEET POTATO——CONTINUED

Start sweet potato from growing shoots or "slips." The leaves look very similar to its relative, morning glory.

With proper storage, they will keep 6 months or more. You can also freeze the cooked roots.

CULTIVARS: Sweet potatoes are divided into dry-flesh and moist-flesh types. The latter are often called "yams," although they are not botanically related to yams. 'Georgia Jet', 'Vardeman', and 'Centennial' are moist-flesh sweet potatoes. Dry-flesh cultivars, which keep better, include 'Orlis' and 'Yellow Jersey'.

SWISS CHARD

Vigorous and easy to grow, a single planting of Swiss chard can provide a full season of fresh greens. Use it like spinach.

BEST CLIMATE AND SITE: Zones 3 and warmer; grow as a winter vegetable in mild areas. Full sun or, in warm areas, partial shade.

IDEAL SOIL CONDITIONS: Not fussy, but prefers rich, well-drained soil; pH 6.0–6.8.

GROWING GUIDELINES: Sow 1–2 weeks before last spring frost. Plant ½ inch (12 mm) deep and firm the soil with the back of a hoe. Thin to stand 8–12 inches (20–30 cm) apart, using thinnings in salads or transplanting them to new beds. Cultivate, and water regularly to keep plants growing strongly.

PEST AND DISEASE PREVENTION: Little troubled by pests. Row covers will deter flea beetles.

COMMON PROBLEMS: Swiss chard is fairly drought-tolerant, but water stress causes tough stems.

DAYS TO MATURITY: 50–60 days; thinnings may be harvested sooner.

HARVESTING AND STORING: Pick large outer leaves by pulling stems from the base with a slight twist. Leave the center to sprout new leaves. Leaves are usually cooked separately from the wide inner rib, which is often steamed and eaten like asparagus. You can freeze leaves as you would spinach.

SPECIAL TIPS: Swiss chard is attractive and has a neat growing habit. Interplant it with flowers for an ornamental and edible border.

CULTIVARS: 'White King', 'Fordhook Giant', 'Rhubarb Chard' (red-stemmed cultivar).

OTHER COMMON NAMES: Leaf beet, silverbeet.

Lycopersicon esculentum   Solanaceae

TOMATO

The large, meaty beefsteak tomato is the classic, old-fashioned garden tomato. Most cultivars need sturdy stakes or cages.

The fruits of yellow currant tomatoes are produced in abundance. Use them whole in salads.

BEST CLIMATE AND SITE: Zones 3 and warmer. Full sun.

IDEAL SOIL CONDITIONS: Fertile, deep, well-drained soil; pH 6.0–7.0.

GROWING GUIDELINES: In long-season areas, tomatoes may be direct-seeded, but they are usually started indoors 5–6 weeks before last spring frost. Set out 24–36 inches (60–90 cm) apart if plants are to be allowed to sprawl, or as close as 15 inches (38 cm) if they are to be staked or caged. Use black plastic mulch to prewarm the soil in short-season areas. Plant tomatoes by burying the stem horizontally, right up to the topmost leaves. New roots will emerge from the buried stem, making a sturdier plant. If the tomato is staked, tie the vine to the stake with soft cloth strips, not wire or string, to avoid damaging the plant. Some gardeners prune tomatoes by pinching out leafy shoots that emerge from leaf axils, but pruning is not necessary. Water tomatoes regularly but do not fertilize until the plant is well established and in full blossom. Then give weak compost tea or fish emulsion. Too much nitrogen will result in lots of foliage but few fruits. Mulch will help retain soil moisture but will also cool the soil, so do not mulch tomatoes until the soil is well warmed.

PEST AND DISEASE PREVENTION: Protect young plants from cutworms with cardboard or metal collars. Handpick tomato hornworms or control them with BT. Folklore states that planting basil with tomatoes helps repel hornworms. To avoid soilborne diseases, do not plant where tomatoes or their relatives, such as eggplant and potato, have grown the previous 2 years. If that is not possible, grow disease-resistant cultivars if you suspect your soil harbors diseases such as Fusarium or Verticillium wilt.

COMMON PROBLEMS: Blossom end rot indicates that the plant is not taking up enough calcium from the soil. If soil calcium levels are adequate, the problem is probably a lack of soil moisture. Keep soil moist but not soggy. Dark brown, circular spots on tomato leaves suggest fungal disease or blight, which can be a serious problem in prolonged humid weather. Provide good air circulation and do not disturb plants when they are wet.

DAYS TO MATURITY: 52–90 frost-free days from transplant. Small-fruited cultivars mature fastest.

HARVESTING AND STORING: Pick the fruit when it is evenly colored but still firm. Can, freeze, or dry excess tomatoes. Cover tomato plants to protect them through a light early frost, but harvest all fruit when more severe frost threatens. Completely green tomatoes will not ripen; pickle them or make into chutney. Blemish-free fruit that has begun to whiten or turn color will continue to ripen: Wrap each tomato in tissue or newspaper and keep in a spot that stays above 55°F (13°C); check often

Low-acid tomatoes like 'Orange Boy' are a good choice for those who prefer a milder, sweeter tomato.

Cherry tomatoes are prolific producers available in many colors to add sparkle to summer salads.

and discard any fruit that develops bad spots.

SPECIAL TIPS: Tomato cultivars are usually described as "indeterminate" or "determinate." Indeterminates continue to grow and set fruit all season, and usually require a sturdy cage or stake. Determinates stop growing and flowering when they reach a certain height. They are often earlier to mature and may not require caging, but they also produce fruit for shorter periods than indeterminates. The letters that are often given after a cultivar's name indicate resistance to the following diseases or pests: V—Verticillium wilt, F—Fusarium wilt, N—Nematodes, T—Tobacco mosaic virus. Where space is limited, plant tall-growing tomatoes at the back of the flower garden or tie them to a wire fence.

RELATED PLANTS:

BEEFSTEAK TOMATO (*L. esculentum*): Large fruit, up to 2 pounds (1 kg) and more, with meaty flesh and often a thick central core. Most are indeterminate. Cultivars include 'Burpee's Supersteak Hybrid', 'Beefmaster', and many heirloom cultivars such as 'Giant Belgium' and 'Pink Ponderosa'.

CHERRY TOMATO (*L. esculentum*): Prolific bearers of fruit that is usually about 1 inch (2.5 cm) in diameter, often borne in grape-like clusters. Cultivars include 'Sweet 100' (red, indeterminate), 'Gold Nugget' (yellow-orange, determinate), and 'Green Grape' (green-yellow, determinate).

CURRANT TOMATO (*L. pimpinellifolium*): Has tiny, currant-sized fruit. Red and yellow variants are available. The vigorous, indeterminate vines will need cages or stakes.

LOW-ACID TOMATO (*L. esculentum*): Reduced acidity usually means milder flavor, with sweet undertones. Yellow-skinned cultivars are often, but not always, low in acid. Try 'Pink Girl' (pink skin, indeterminate), 'Lemon Boy' (yellow skin, indeterminate), 'Jetstar' (red skin, determinate), and 'Orange Boy' (orange skin, indeterminate).

PASTE TOMATO (*L. esculentum*): Meatier than other tomatoes, with fewer seeds and less juice. Most cultivars are determinate, and many have concentrated fruit-set, with up to 90 percent of fruit ripening at once. Cultivars include 'Roma' (small, pear-shaped), 'Milano' (early maturity, plum-shaped), and 'Royal Chico' (large, flat-sided, plum-shaped).

PATIO TOMATO (*L. esculentum*): Determinate cultivars bred for growing in containers, but also good for small gardens. Cultivars include 'Patio' and 'Better Bush'.

PEAR TOMATO (*L. esculentum* var. *pyriforme*): The fruits look like tiny pears. Red and yellow variants are available. The vigorous, indeterminate vines will need staking or caging.

SLICING TOMATO (*L. esculentum*): Grown primarily for fresh use. Scores of cultivars are available, in a

TURNIP

Paste tomatoes have fewer seeds and less juice than slicing tomatoes, making them ideal for homemade soups and sauces.

Grown both for its tender leaves and for its crisp root, turnip is at its best in cool weather.

wide range of colors, growth habits, maturity dates, and disease resistance. Among the most popular are 'Early Girl' (small red fruit, early-ripening), 'Big Boy' (large red fruit), 'Celebrity' (large red fruit, very disease-resistant), and 'Jubilee' (large yellow-orange fruit).

Pear tomatoes set abundant fruits that are just the right size for salads and pickling.

BEST CLIMATE AND SITE: All Zones; grow as a winter vegetable in mild climates. Full sun.

IDEAL SOIL CONDITIONS: Loose, deep, humusy soil; pH 5.5–6.8.

GROWING GUIDELINES: Sow ½ inch (12 mm) deep 4–6 weeks before last spring frost. Thin to stand 3 inches (7.5 cm) apart; use thinnings as fresh or cooked greens. Water regularly for fast growth in spring, since turnip bolts in hot weather. Sow fall crop 8–10 weeks before first fall frost.

PEST AND DISEASE PREVENTION: Use row covers to protect from flea beetles. Reduce root maggot damage, especially in spring crops, by not planting where other root crops have grown the previous year.

COMMON PROBLEMS: Turnips that mature in hot weather may be fibrous or strong-flavored. For spring crops, choose quick-maturing cultivars.

DAYS TO MATURITY: 35–60 days. Turnips will withstand light frost.

HARVESTING AND STORING: Pull turnips as needed, as 1-inch (2.5 cm) wide "babies" up to 3- to 4-inch (8–10 cm) roots. Larger turnips may be woody. Turnips are more tender than other root vegetables and are damaged by hard frost. Store roots in damp sand or sawdust in a cool place; freeze leaves as you would spinach.

CULTIVARS: 'Tokyo Cross Hybrid', 'Purple-Top White Globe', 'Seven Top' (grown for greens), 'Golden Ball' (yellow flesh).

| *Brassica rapa,* Rapifera group | Cruciferae | *Proboscidea louisianica* | Martyniaceae |

TYFON

UNICORN FLOWER

A mild-flavored cross between turnip and cabbage, tyfon is a quick-growing, hardy green that survives low temperatures.

BEST CLIMATE AND SITE: All Zones; grow as a winter vegetable in mild areas. Full sun or partial shade.

IDEAL SOIL CONDITIONS: Not fussy, but prefers rich, well-limed soil; pH 6.0–7.0.

GROWING GUIDELINES: Sow ½ inch (12 mm) deep 4–6 weeks before last spring frost, 8–10 weeks before first fall frost, or later as a winter vegetable in mild areas. Thin to stand 4–6 inches (10–15 cm) apart; use thinnings as salad or cooking greens. Or grow unthinned if you plan to harvest leaves by shearing. Water regularly to encourage abundant leafy growth.

PEST AND DISEASE PREVENTION: Use row covers to protect plants, especially spring crops, from flea beetles.

COMMON PROBLEMS: Avoid plantings that will mature in hot, dry weather. Like its cabbage-family relatives, tyfon bolts quickly in the heat.

DAYS TO MATURITY: 90 days.

HARVESTING AND STORING: Pick fresh leaves as needed, or harvest by cutting with garden shears or scissors, leaving 1 inch (2.5 cm) above the crown to resprout. In cool weather, a planting should provide several cuttings, about a month apart. Freeze leaves as you would kale or spinach.

SPECIAL TIPS: In most climates, a light row cover will extend harvest well into the winter.

CULTIVARS: Named cultivars not generally available.

OTHER COMMON NAMES: Holland greens.

Often grown as an ornamental, the unicorn flower produces attractive flowers and curiously shaped fruits.

BEST CLIMATE AND SITE: Zones 4 and warmer. Full sun.

IDEAL SOIL CONDITIONS: Well-drained, neutral soil; pH 7.0.

GROWING GUIDELINES: Direct-seed in mild climates; in cool climates, start plants indoors and set out after frost danger is well past. Set plants 3–5 feet (90–150 cm) apart, as they will spread. Water regularly and cultivate or mulch until the large leaves shade out weeds.

PEST AND DISEASE PREVENTION: Little troubled by pests.

COMMON PROBLEMS: If picked too mature, the fruits will be tough and leathery.

DAYS TO MATURITY: 50 or more frost-free days from transplanting.

HARVESTING AND STORING: Pick tender, young fruits when they are 2–3 inches (5–8 cm) long. Pickle as you would cucumbers. Oversized fruits, especially near season's end, may be left on the plant to mature completely. They will dry, become woody, and split open. The resulting "devil's claw" is useful in dried winter arrangements.

RELATED PLANTS: Named cultivars not generally available. *Ibicella lutea,* native to South America, produces similar but smaller fruits.

OTHER COMMON NAMES: Martynia, proboscis flower, ram's horn.

Viola spp.	Violaceae

VIOLET

Cherished for their cheerful spring flowers, violets and their relatives, pansies and violas, are just as sprightly atop a salad.

BEST CLIMATE AND SITE: All Zones; grow as a winter crop in Zones 8 and warmer. Partial shade.

IDEAL SOIL CONDITIONS: Rich, humusy, moist soil; pH 6.0–8.0.

GROWING GUIDELINES: Start indoors in spring, 12–14 weeks before last frost; or sow the previous summer and overwinter the plants in a cold frame for setting out in spring. Space plants 6–8 inches (15–20 cm) apart and cultivate or mulch. Water regularly and pinch off faded flowers for extended bloom. Some *Viola* species are perennial; others are treated as biennials or annuals.

PEST AND DISEASE PREVENTION: Trap slugs in shallow pans of beer set flush with the soil surface.

COMMON PROBLEMS: Plants stop blooming in hot weather. Many annual cultivars will bloom again in fall if kept watered.

DAYS TO MATURITY: Generally 120–150 days for spring-grown violas and johnny-jump-ups. Fall-grown pansies will bloom the following spring. Sweet violets are spring-blooming perennials.

HARVESTING AND STORING: Pick fresh blossoms as needed in early morning. Refrigerate immediately in a covered container. Candy blossoms or dry to add to herbal teas.

RELATED PLANTS: Johnny-jump-up (*V. tricolor*); sweet violet or English violet (*V. odorata*); viola (*V. cornuta*); pansy (*V. x wittrockiana*). Pansy cultivars include 'Jolly Joker', 'Swiss Giants', and 'Toyland'.

Nasturtium officinale	Cruciferae

WATERCRESS

At its best in fall and early spring, watercress is a tasty and healthful addition to salads and sandwiches.

BEST CLIMATE AND SITE: Zones 3 and warmer; grow as a winter vegetable in warmer areas or indoors in pots almost anywhere. Full sun or partial shade.

IDEAL SOIL CONDITIONS: Wet, humusy, well-limed soil, preferably at the edge of a stream or stream-fed pond; pH 6.0–7.0.

GROWING GUIDELINES: Start indoors 4–8 weeks before the last spring frost, or sprinkle seeds as thinly as possible where they are to grow, about a month before the last frost, pressing them into the soil. Fill gaps in planting by breaking off pieces of stem and pressing them into the soil, where they will root easily. Gardeners without access to constantly moist soil may grow watercress in pots, indoors or out. Set pots in pans of water and change the water daily.

PEST AND DISEASE PREVENTION: Little troubled by pests and diseases.

DAYS TO MATURITY: 120–150 days. In short-season areas, a newly established watercress bed may not produce a harvest the first year.

HARVESTING AND STORING: Harvest leaves as needed, midfall through early spring, including winter where climate permits. Flavor deteriorates during flowering.

CULTIVARS: Named cultivars not generally available.

USDA
PLANT HARDINESS ZONE MAP

The map that follows shows the United States and Canada divided into 10 zones. Each zone is based on a 10°F (5.6°C) difference in average annual minimum temperature. Some areas are considered too high in elevation for plant cultivation and so are not assigned to any zone. There are also island zones that are warmer or cooler than surrounding areas because of differences in elevation; they have been given a zone different from the surrounding areas. Many large urban areas are in a warmer zone than the surrounding land.

Plants grow best within an optimum range of temperatures. The range may be wide for some species, and narrow for others. Plants also differ in their ability to survive frost and their sun or shade requirements.

The zone ratings indicate conditions where designated plants will grow well, and not merely survive. Refer to the map to find out which zone you are in. In the "Plant by Plant Guide," starting on page 78, you'll find recommendations for the plants that grow best in your zone.

Many plants may survive in zones warmer or colder than their recommended zone range. Remember that other factors, including wind, soil type, soil moisture and drainage capability, humidity, snow, and winter sunshine, may have a great effect on growth.

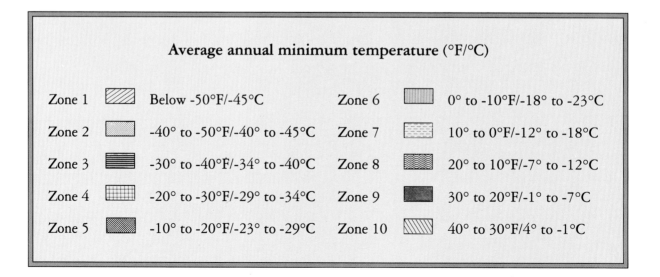

Average annual minimum temperature (°F/°C)

Zone 1	Below -50°F/-45°C	
Zone 2	-40° to -50°F/-40° to -45°C	
Zone 3	-30° to -40°F/-34° to -40°C	
Zone 4	-20° to -30°F/-29° to -34°C	
Zone 5	-10° to -20°F/-23° to -29°C	
Zone 6	0° to -10°F/-18° to -23°C	
Zone 7	10° to 0°F/-12° to -18°C	
Zone 8	20° to 10°F/-7° to -12°C	
Zone 9	30° to 20°F/-1° to -7°C	
Zone 10	40° to 30°F/4° to -1°C	

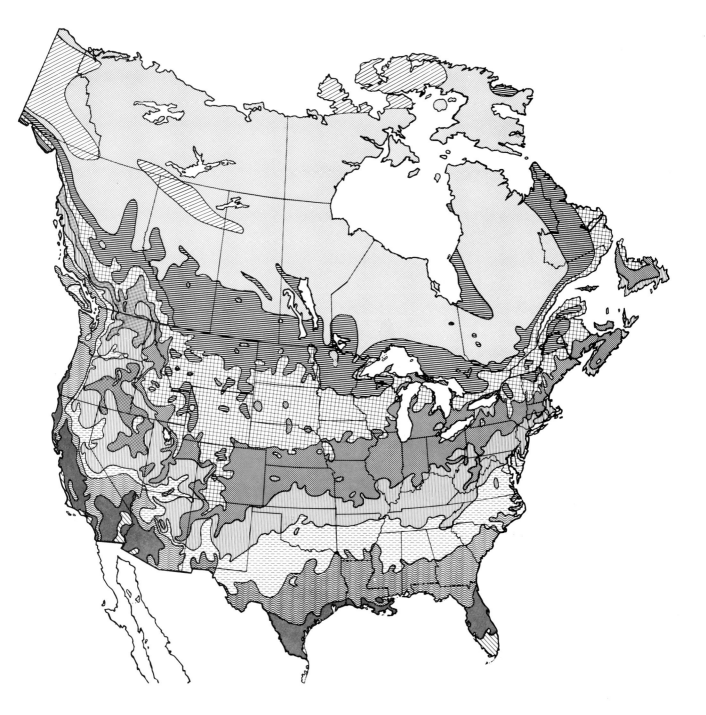

INDEX

ACKNOWLEDGMENTS

Weldon Russell would like to thank the following Australian companies and people for their assistance in the production of this book: June Bland; Colonial Cottage Nursery, Kenthurst; Common Scents Cottage, Dural; Duane Norris Garden Designers, Woollahra; Fragrant Garden, Erina; Henry Doubleday Research Association; Honeysuckle Garden Centre, Double Bay; Russell Lee; Di McDonald; McNaturals Nursery, Hazelbrook; Catherine Wallace.

Photo Credits

Ardea London: opposite contents, pages 123 (photographer A. P. Paterson) and 130 (left, photographer Don Hadden).

Auscape/Jacana: pages 15 (right), 22, 35 (top left and center, photographer John McCammon).

A–Z Botanical: pages 103 (left), 113 (left), 115 (left), and 139 (right).

Heather Angel: pages 8, 16 (right), 92 (left), 95 (right), and 114 (right).

Bruce Coleman Ltd: pages 83 (left), 88 (right), 116 (right), 136 (right, photographer Eric Crichton), and 152 (right, photographer Michel Viard).

Peter Devaus: page 98 (left).

Thomas Eltzroth: pages 11 (bottom left), 15 (left), 32 (top), 35 (top right), 40, 55 (top), 60 (top left and bottom right), 64 (top left and top right), 65 (top and bottom), 77 (bottom), 84 (right), 89 (left), 98 (top right and bottom right), 100 (left), 101 (right), 108 (left), 121 (left and right), 124 (bottom left), 127 (right), 133 (left), 139 (left), 144 (right), 146 (top right), 149 (left), 150 (top left), 151 (left), and 152 (left).

Derek Fell: back cover (center), pages 29 (bottom), 47, 58 (bottom), 59 (bottom right), 70, 71, 72 (bottom center), 83 (right), 103 (right), 105 (left), 109 (right), 138 (left), 141 (right), and 146 (left).

Lee Foster: pages 19 (top and bottom), 28, 42, 46, 58 (top), and 73 (top).

Rowan Fotheringham (stylist Karen Byak): front cover.

The Garden Picture Library: photographer Linda Burgess: endpapers and copyright page; photographer Tommy Candler: page 26 (bottom); photographer Brian Carter: page 74 (bottom center); photographer Geoff Dann: page 53 (top); photographer Robert Estall: back cover (bottom), pages 72 (bottom left) and 75; photographer John Glover: opposite title page and page 73 (bottom center); photographer Jerry Pavia: page 26 (top); photographer Joanne Pavia: page 21 (top); photographer Ron Sutherland: page 77 (top); photographer Brigitte Thomas: page 18.

Harry Smith Collection: pages 33 (bottom right), 106 (right), 109 (left), 114 (left), 118 (right), 134 (top left and bottom left), 141 (left), and 147 (right).

International Photographic Library: pages 32 (bottom) and 143 (left).

Stirling Macoboy: pages 85 (bottom right), 97 (left), and 136 (left).

Cheryl Maddocks: pages 80 (right), 81 (right), 86 (right), 120 (top left), 138 (right), and 144 (left).

Jerry Pavia: page 52.

Joanne Pavia: half title page, contents page (center right), pages 27 (top), and 52.

Photos Horticultural: back cover (top), pages 11 (bottom right), 16 (left), 29 (top), 33 (top), 34, 50 (top), 51 (top), 54, 55 (bottom left and bottom right), 59 (bottom left), 68, 69, 72 (bottom right), 73 (bottom left and bottom right), 74 (bottom left and bottom right), 76, 78, 96 (right), 99 (left), 100 (right), 107 (left and right), 122 (left), 125 (left), 127 (left), and 149 (right).

Rodale Stock Images: pages 14, 17, 21 (bottom), 30, 33 (bottom left), 45, 56 (bottom), 57 (left and right), 66, 67 (top), 85 (left), 86 (left), 87 (right), 88 (bottom left), 90 (right), 91 (right), 93 (left), 102 (right), 104 (right), 112 (left and right), 117 (left), 124 (right), 126 (top right and bottom right), 129 (top right and bottom right), 145 (left), 146 (bottom right), and 150 (right).

Tony Rodd: pages 53 (bottom), 85 (top right), 95 (left), and 118 (left).

Lorna Rose: page 12.

Weldon Trannies: contents page (bottom left).

All other photographs by David Wallace.